THE YEAR OF THE ROBOT

Wayne Chen

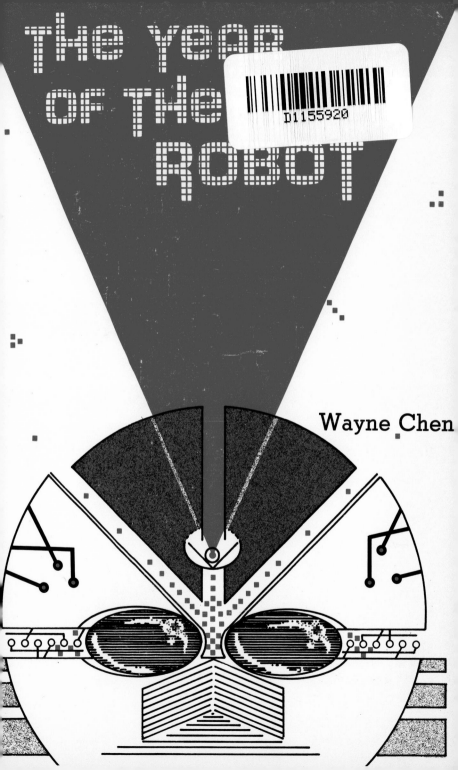

The Year of the Robot

The Year of the Robot

Wayne Chen

dilithium Press
Beaverton, Oregon

DISCLAIMER

Any similarity between the reader's traits and behavior and those of the robots, as described in this book, is unintentional, but probably more than merely accidental.

NMU LIBRARY

ISBN: 0-918398-50-9
Library of Congress catalog card number: 80-52890

10	9	8	7	6	5	4	3	2	1

Printed in the United States of America.

dilithium Press
P.O. Box 606
Beaverton, Oregon 97075

PREFACE

Much has been written about machines that have human intelligence. Almost all of these machines are computers in some form.

We deal in this book with a special class of machines that are *not* computers. However, such a machine can be related to the computer in that *either* it is a component, as an amplifier or a servomechanism, of a computer, *or* the computer is a component of this machine, as a missile guidance system or a process control system or any complex feedback control system. Such a machine is found to possess human traits that are *intellectual* in nature, and hence is called an "intellectual robot" or a "robot."

This book provides an unusual perspective about technology and humanity. It begins with a study of the human traits of the robot, and then applies them to the interpretation of human behavior, social phenomena and our democratic institution.

In the least, this book represents an interesting exercise in relating technology to humanity. In addition to its role of providing wondrous products to serve human beings, technology in the person of this robot is encroaching upon the turfs of religion, morality and philosophy, to teach us how to behave. Although the underlying principle of the robot is scientifically valid, this book is also meant to be humorous, entertaining and to some degree even evangelical. (Who in his right mind wants to worship an almighty robot?)

This book consists of two parts and is centered about the intellectual robot:

Part I, entitled "The Intellectual Robot," is a philosophical dissertation about the human traits of the robot and their application to the interpretation of human behavior, social phenomena and our democratic institution.

Part II, entitled "The Robotosyncrasies," is a novel.[1] Since the robot is really the creation of numerous contributions in an evolutionary process throughout the years, the author uses a fictional approach in the book in the form of a love story about the fictitious inventor and his journalistic sweetheart. A movie-like scenario landscaped with the social flora (international settings, space exploration, women's lib, protest movements, etc.) of our time is used to show the social implications of the robot and the pervasiveness of technology.

Part II was published as a novel in 1976. As the author was contemplating an article on "The Intellectual Robot," the publisher persuaded him to write instead a new version (Part I here) and to publish the new writings together with his earlier novel in this combined volume.

What has been told about the robot is a *true* story. With two parts of the book together, this volume may be considered as a *biography* of the robot.

If you are an *electrical engineering student* or an *electrical engineer,* did it ever occur to you that a feedback amplifier or a servomechanism, as you have learned how to design, has a "pulsating heartbeat" that resonates with the rhythms of your life and those of our social institutions? Discover how to apply your feedback theory to your own conduct as well as to the conducts of our society, our nation and the world!

If you are a *student* or a *professional* in *engineering, mathematics or sciences,* have you ever suspected that a closed-loop system, a feedback control system, a process control system or a missile guidance system has a "soul" that sustains this machine and also helps explain our many social phenomena which are often considered inexplicable? Let us explore the underlying principles of these systems and discover for oursleves the intricacies of our society.

Individuals, as well as *leaders* in *social movements, business and industry, governments* and *the world,* you may wish to see and understand your own traits and conducts and those of your social institutions, as well as their pitfalls and future trends, as reflected through the robot's eye. In this book,

---

[1] In portions of Part II, the author misused the word "intelligent" where the word "intellectual" is actually meant.

"reflectiveness" is discovered to be the open-sesame trait, and through some cause-and-effect relations also helps spawn other desirable human traits and social conducts. What is amazing is that these social behavior patterns are dictated by the Law of Nature (i.e., some basic knowledge of physics as the underlying principle of the intellectual robot) and are not arbitrarily suggested or contrived.

For the general readership, while technology may seem to be "cold" and "impersonal," this book has now humanized it in a philosophical dissertation (Part I) and in a novel about love and conflict (Part II). In reading the novel, the reader may omit Chapter 6 of Part II and substitute it with Chapter 2 of Part I. The latter is a simpler version of the former with no mathematics and no derivations.

The author apologizes for using some simple mathematics and for using an unusual format in this book. The use of simple mathematics is necessary to establish the technical validity of the intellectual robot, and this use is deliberately confined to one place (Chapter 6 of Part II) so that the reader may read it, omit it, or substitute it with a simpler version, without reading interruption. Although the material is true on a scientific basis, it will look and sound "unreal" and "unbelievable" even in the format of a conventional essay or novel. For this reason, the author just chooses a format to put his ideas across, while still preserving the scientific basis of his material.

CONTENTS

Part I

The Intellectual Robot

INTRODUCTION

What is so unique and special about the human being? The human being is, metaphorically speaking, endowed with the hand, brain, and heart.

The *hand* symbolizes dexterity and skills of the highest order. The *brain* teaches us how to think cleverly and achieve the ingenious. And the *heart*, in an ethical, moral, religious and philosophical sense (but, oddly enough, not physiologically) guides us intellectually and compassionately.

We live in the age of technology. We have smart bombs which can zero in on targets thousands of miles away. We have satellites for remote-sensing or spying on the earth and identifying resources or objects, much like finding a needle in a hay stack. We use automated machinery to finish products, step by step, in an assembly line. Mechanical robots can be readily used to substitute for humans in a hostile environment; in repairing the heat-exchanger in a nuclear reactor or in performing difficult, delicate and precise operations. Computers are used in a variety of ways as a substitute for human intelligence. All these are "intelligent" machines that are literally endowed with the human "hand" and "brain." They are often referred to as "robots" or the "intelligent robots."

We shall examine another class of machines that have a number of human traits; how to be broadminded and fair, to be responsive without overreacting, to be visionary and deny self-doubt, to be resolute and reliable, etc. In other words, a class of machines with a *heart*. Lacking appropriate words to describe it, we shall also call it a robot. Here we have an "intellectual robot."

It is also amazing to discover that the intellectual robot has a master ("open sesame") human trait; many other traits come as a consequence and can be readily derived from the master trait through a cause-and-effect relationsihip. Is this cause-and-effect relationship among the robot's human traits also applicable to human beings? This is a provocative concept that can have an impact on our religious, moral and philosophical thinking.

The "intellectual robot" is also found to be a microcosm of our society, and of our political system. It is fascinating that the traits of the robot can be readily used to explain some of our social phenomena which are often considered inexplicable. The robot explains them, rightly or wrongly, with new perspectives. Through the robot's eye, we can learn how our democratic process works.

A comparison of the brief recent histories of machines and human beings also reveals that a robotic type of revolution has been under way in the human society.

1. IN THE BEGINNING

The "intellectual robot" and its human traits were discussed in the author's book *The Robotosyncrasies* under the pen name of Wayne Hawaii.[1].

All robotosyncrasies, the human traits of the robot, are derived from a simple system model which itself is a human trait in *reflectiveness*. The simple system model requires that the machine always compare what it gets with what it wants; the machine "reflects" on its own performance and goal. In other words, "reflectiveness" is responsible for a host of human traits in the robot.

Hawaii presupposes that the robot is really the creation of numerous contributors in an evolutionary process through the years. For this reason, he uses a fictional approach in the book in the form of a love story about the fictitious inventor of the robot and his journalistic sweetheart. This is intended to enhance the reader's interest, and to present an unusual perspective about technology and the human element. The scenario and its backdrop — including the Apollo 11 moon mission which was certainly made possible by the robot as described — were designed to show the "ubiquitousness" and "pervasiveness" of the technology identified with this intellectual robot as well as its impacts upon our daily life. And while the robot is found to have human traits, its social implications are implicitly told throughout the plot of the book.

---

[1] Wayne Hawaii, *The Robosyncrasies*, Carlton Press, 1976, which is now published as Part II of this book.

Some excerpts of two book reviews that appeared in professional society publications, *Mechanical Engineering*[2] and ASEE/IEEE *Newsletter*[3] are cited below to substantiate[4] our comparison between the robot and human beings and various discussions. For example, one reviewer comments about the parallelism between technology and humanity:

"The engineer-author's unusually sensitive discernment of parallels between technology and humanity brought him to point out and to make plausible heretofore unrecognized similarities between the traits of people and machines. Each *robotosyncrasy* he describes is shown to be the direct parallel of a human characteristic that some of us may possess, and all of us would do well to try to acquire. He lists these traits as even-temperedness, broadmindedness, agility, stability, vision, resoluteness, and reliability. And he first establishes each parallel by simple, intuitive explanation and then mathematically derives each robotosyncrasy or human characteristic from a simple system model. I found both fascinating although the second kind, the mathematical derivations, can safely be skipped without losing the continuity of the novel."[2]

The technology of the robot which we discuss is not an isolated example of exotic and esoteric machines. Instead, it can be found *everywhere* in our homes, offices, factories, research laboratories, and the arsenals of the armed forces. One of the reviewers comments about the "ubiquitousness"

[2] George N. Sandor, Book Review on "The Robotosyncrasies," *Mechanical Engineering*, Published by the American Society of Mechanical Engineers (ASME), August 1977, pp. 111-112.

[3] Matthew Kabrisky, Book Review on "Robotosyncrasies," *Newsletter* Published Jointly by The American Society for Engineering Education (ASEE) Electrical Engineering Division and The Institute of Electrical and Electronics Engineers (IEEE) Education Group, February 1977, p. 12.

[4] While most of the contemporary writings are based upon existing knowledge, the author in this book has literally "created" or "invented" the human traits of the robot (which he calls the robotosyncrasies) as well as their unusual interpretations in the social and political contexts. Normally, the author would never consider citing book reviews of his early work. However he is using some book review statements here to ensure the scientific validity of this robot and the associated "invented" body of knowledge, and to provide two important historical milestones in the creation of the robot, which he believes will be of considerable interest to the reader.

and pervasiveness" of this (negative feedback) technology of
the robot:

> "The novel supposes that the entire technology of negative
> feedback as well as *all* of its consequences including ser-
> vos, broadband operational amplifiers, missile guidance
> systems, etc., was invented and completely exploited by
> one man. Further, that this happened suddenly and very
> recently since the man is in his late 30's at about the time
> of the Armstrong-Aldrin Apollo 11 flight to the moon (an
> event in this book). The hero is suitably feted and finan-
> cially rewarded as might well happen if the story were ac-
> tually true. This fictional hypothesis serves well to em-
> phasize the *ubiquitous, pervasive,* and incredibly impor-
> tant consequences of negative feedback technology which
> in everyday life is totally unrecognized by the average
> man in the street and relegated to the reverence associated
> with just another junior level course by EE's (electrical
> engineers)."[5]

The author's presupposition that the robot was created by
numerous contributors in an evolutionary process was actually
challenged and also substantiated (by reading two reviews
together) by the reviews which point out, very significantly,
some *historical milestones* for the creation of the intellectual
robot:

> "The author's rare insight is attested to by the fact that
> robotosyncrasies, or anything like them, have never been
> pointed out before, although stabilized feedback control-
> lers, albeit not called by this name but referred to as
> 'governors' or 'regulators,' similar to the author's 'robot,'
> have been around in mechanical engineering for over two
> centuries — ever since James Watt connected his cen-
> trifugal fly-ball governor to the throttle valve of his steam
> engine by way of an adjustable turnbuckle. The turn-
> buckle provided the speed setting or 'command signal,'
> while the fly-balls generated the feedback signal by sens-
> ing the engine speed. And today's sophisticated
> mechanical engineering control systems comprising not
> only mechanical but also thermal, fluidic, hydraulic,

---

[5] Kabrisky, *loc. cit.*

pneumatic, aerodynamic, as well as electric-electronic elements and subsystems, certainly display all the 'robotosyncrasies' so vividly described in Wayne Hawaii's fascinating book."[6]

"The principle of negative feedback was invented by Harry S. Black one night on fogbound Weehauken Ferry boat (see H.S. Black 'Stabilized Feedback Amplifiers.' *B.S.T.J.* or *Electrical Engineering*, January 1934). He in fact patented negative feedback (assigned to the Bell Telephone Labs, U.S. Patent No. 2,102,671) but he never got anything out of it except his normal salary at Bell Labs plus one dollar and quite a bit of obscurity. It is well known to EE's (electrical engineers) that the subsequent exploitation and myriad applications of feedback have been the result of the ideas and work of thousands of very clever people over a period of four decades."[7]

While Hawaii's book derives or demonstrates the robotosyncrasies with simple mathematics, this writing will only cite and explain the robotosyncracies as true "scientific facts" with *no mathematics* (and no derivations) and will also study their cause-and-effect relations as well as social implications.

In Hawaii's book, sections of an advanced engineering book also by the author under his legal name Wayne Chen[8] are included as Appendices A and B to back up its simple explanations and discussions in establishing the robotosyncrasies. Readers who are interested in how these robotosyncrasies are derived are referred to Hawaii and/or Chen.

---

[6] Sandor, *loc. cit.*

[7] Kabrisky, *loc. cit.*

[8] Wayne H. Chen, *The Analysis of Linear Systems*, McGraw-Hill Book Co., 1963. Portions of Chen's book are included as appendices of Part II of this book.

Commentary:

2. THE ROBOTOSYNCRASIES[1]

What is amazing about these robotosyncrasies is that *most* of them can be readily and simplistically derived from a very simple algebraic equation $G* = G/(1+GH)$. This is why Hawaii can handle it with simple mathematics. However, *some* robotosyncracies need more sophisticated treatment (Chen).

We shall only cite and explain the robotosyncrasies as true scienfitic facts with no mathematics and no derivations in this section.

The Robot

Let us examine a class of machines that for a given *command* (or *input*), we wish to produce an *effect* (or *response*) that obeys and follows the command (or input). In other words, we want the command and effect to be "identical" in shape.

For example, we may give a "command" for the control of gun fire or machine speed or manufacturing process, and wish to produce an "effect" or "result" that follows our given command. Or for another example, we may provide an "input" to the amplifier in our hi fi set, and wish to produce an "output" or "response" which duplicates exactly the input in order to preserve the quality of high fidelity.

A "crude machine" designed for this purpose is schematically shown by a G-box in Fig. 1.

A "robot" is a modified form of a crude machine. It continuously compares what it gets (the effect, y) with what it wants (the command, x). As shown in Fig. 2a, it has a "feedback loop" that brings back the effect to be compared with the command, and has a "comparator" that compares x and y and feeds its difference to the crude machine G.

---

[1] Those readers who have an aversion to scientific terminology in Chapter 2 may skip this chapter and read the same message in a condensed narrative form in *Appendix I-A* (PART I of this book) instead.

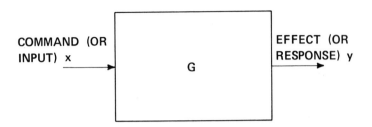

Fig. 1

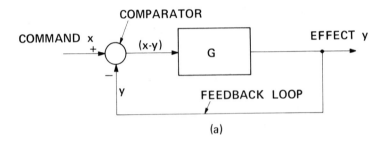

(a)

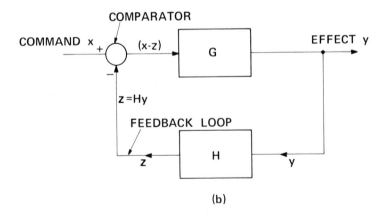

(b)

Fig. 2

However, we sometimes run into trouble when we try to compare the effect with the command. For example, if the robot is designed to control the speed of a machine, the command is often an *electrical signal* measured in volts and the effect is a *speed* measured in revolutions per minute. We really cannot compare x and y, and make sense of what quantity is its difference. It is almost like comparing "10 monkeys" and "5 quarts of milk," and asking what quantity is their difference. It makes no sense. In order to compare two quantities, they must be quantities of the *same* physical category or unit.

A more generalized form of a "robot" is shown in Fig. 2b, where for example the command is an electrical signal in volts and the effect is a speed in r.p.m. In order to compare them, we feed y into a transducer H, and obtain a quantity z=Hy (where z is related to y, and is equal to the product of H and y) as another electrical signal in volts. Now, we can compare x and z, *both* electrical signals in volts, and feed their difference into the crude machine G. In reality, the H-box also has the additional responsibility of improving the performance of the robot.

A *robot* is now defined to be a machine that continuously compares what it gets (the effect, y, or its related quantity, z) with what it wants (the command, x) as shown in Fig. 2b or as its special case in Fig. 2a.

For ready reference, we shall represent a robot schematically as a G*-box as in Fig. 3a or 3b or with the legend in Fig. 3c, while we shall represent a crude machine schematically as a G-box in Fig. 1 or Fig. 4a or the legend in Fig. 4b.

The Gain Curve

When we hit a piano key, we hear an *elementary signal* of a certain frequency f. The *frequency* f of an elementary signal is the number of vibrations (or cycles) per second. The name "hertz" (hz) is given to "cycles per second." For example, when we hit the highest C key on the piano, it produces an elementary signal of a frequency f=4,000 hz. (This means that the thin wire that produces the highest C on the piano actually vibrates back and forth 4,000 times a second).

As an example, we may have elementary signals of these frequencies:

$$f_1 = 10 \text{ hz}$$
$$f_2 = 35 \text{ hz}$$
$$f_3 = 10,000 \text{ hz}$$
$$f_4 = 1,000,000 \text{ hz}$$

For a crude machine or a robot, it has a *gain curve* (labeled as a G-curve for a crude machine, or a G*-curve for a robot) as shown in Fig. 5. The gain curve is really a "credential" or "portait" of the machine in its ability to deal with elementary

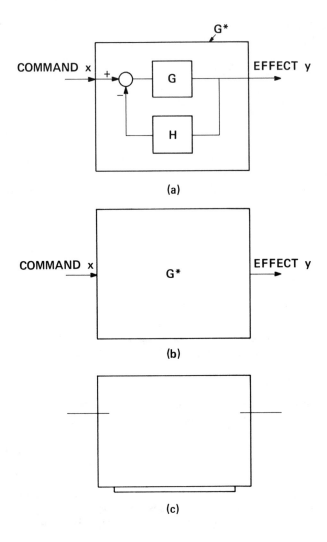

Fig. 3

signals of different frequencies. In the example of the gain curve in Fig. 5, it says that

1) An elementary signal of the frequency f_1 ($f_1=500$ hz) whose magnitude is x (the magnitude is a measure of its loudness as a sound in this example) is used as the input

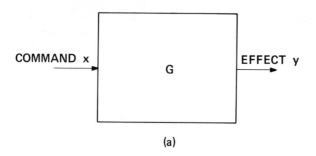

(a)

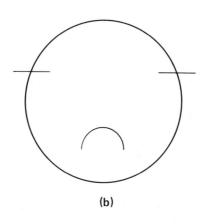

(b)

Fig. 4

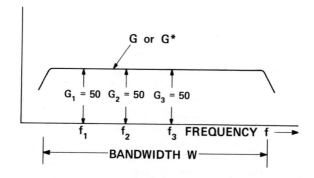

Fig. 5

to the machine. We measure its output magnitude y and have the *gain* (or *amplification factor* or *vitality factor*) of the machinery at the frequency f_1 as

$$\frac{y}{x} = G_1 = 50 \text{ (height of the curve at } f_1)$$

It means that the machine *amplifies* the elementary signal at the frequency f_1 50 times. (For an input magnitude $x=1$, the output magnitude will be $y=G_1 x=50$; and the output y is 50 times as large as the input.)

2) At the frequency f_2 ($f_2=5,000$ hz),

$$\frac{y}{x} = G_2 = 50 \text{ (height of the curve at } f_2)$$

This means that the machine amplifies the elementary signal at this frequency 50 times.

3) At the frequency f_3 ($f_3=50,000$ hz),

$$\frac{y}{x} = G_3 = 50 \text{ (height of the curve at } f_3)$$

This means that the machine amplifies the elementary signal at this frequency 50 times.

We note here that at the frequencies f_1, f_2, f_3:

$$G_1 = G_2 = G_3 = 50$$

And it means that the machine with the gain curve in Fig. 5 is *uniformly fair* to signals of all frequencies f_1, f_2, f_3 . . . in the sense that it treats them equally. It can be said to be *even-tempered* because of its even disposition toward signals in a range of frequencies. The behavior of this machine is highly *predicatble.*

Again using the gain curve in Fig. 5, we shall call the range of frequencies in which the gain factor G remains almost at the highest, G= 50 in this case, the *bandwidth* W of this machine. It begins to drop rather rapidly outside this range. The bandwidth is the "breadth of a robot's mind" to accommodate elementary signals or commands (which are complex signals). In other words, it is the mindwidth of the machine.

Not all gain curves are like the one in Fig. 5. For example, we may have a machine with a "gain curve" as given in Fig. 6. In this case, we note that the machine amplifies the elementary signal:

1) At the frequency f_1,

$$2 \text{ times } (G_1 = 2)$$

2) At the frequency f_2,

$$5 \text{ times } (G_2 = 5)$$

3) At the frequency f_3,

$$10 \text{ times } (G_3 = 10)$$

It means that this machine is *uneven* or *unfair* due to its very different treatments for signals of different frequencies. It could be called *temperamental, ill-tempered, impulsive* or *capricious*. In other words, we have an Ivan the Terrible among our machines.

We shall next discuss the robotosyncrasies or human traits of the robot *without any mathematics*. (For derivations or demonstrations, see Hawaii or Chen.)

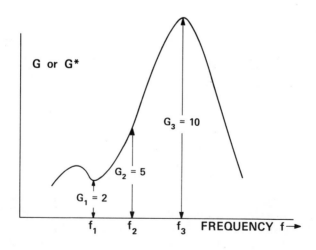

Fig. 6

Robotosyncrasy 1: Broadminded

A robot is much more "broad-minded" that its crude machine.

For a gain curve (the G curve) of a crude machine as shown in Fig. 7a, the gain curve, (the G* curve) of a robot that is made of this crude machine and a feedback structure, can be readily obtained and is shown in Fig. 7b.

We note that the robot has a much broader bandwidth W_2 than its crude machine's bandwidth W_1. In other words, the robot has a broader mind to accommodate a *larger range* (or a *greater constituency*) of signal frequencies.

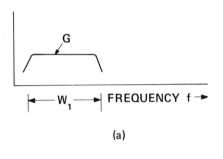

(a)

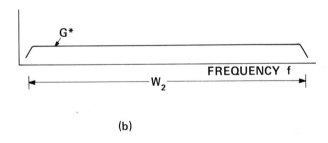

(b)

Fig. 7

Robotosyncrasy 2: Even-Tempered and Fair

A robot is much more "even-tempered" than its crude machine. In other words, it is more "fair" to signals of different frequencies in that it treats them more equally.

Let us assume that the gain curve (the G curve) of an ill-tempered crude machine is shown in Fig. 8a. Here we note that the machine amplifies the elementary signal at the frequency f_1 35 times ($G_1 = 35$) while it amplifies the elementary signal of the frequency f_2 100 times ($G_2 = 100$). Apparently, it treats signals of different frequencies rather *unevenly*: it favors signal f_2 to signal f_1 by a factor of almost 3 ($100 \div 35 = 2.85$).

Now for the G curve of a crude machine as shown in Fig. 8a, we can readily obtain the G* curve of its associated robot (made of this crude machine and a feedback structure) as shown in Fig. 8b.

Here we note that for the same two frequencies f_1 and f_2 the robot amplifies the elementary signal at f_1 8.4 times ($G_1{}^* = 8.4$) while it amplifies the elementary signal at f_2 10 times ($G_1{}^* = 10$). Apparently, it treats signals of different frequencies more *evenly*: it favors signal f_2 to signal f_1 by a factor of only 1.19 ($10 \div 8.4 = 1.19$).

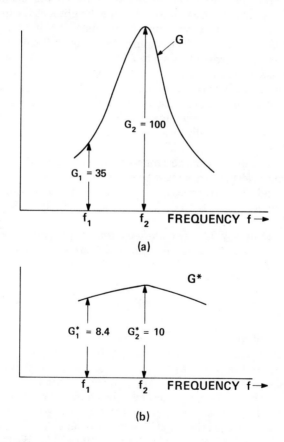

(a)

(b)

Fig. 8

Robotosyncrasy 3: Agile and Responsive

A robot is much more "agile" (faster) in responding to a command than its crude machine. In other words, it is more "responsive" to the demand or goal that it has been asked to achieve.

Before we attempt to discuss this Robotosyncrasy, let us introduce Robotosyncrasy 3A as a basic characteristic or trait of the robot.

When we give a "command" to a machine, we really mean to "set its goal." For example, look at Fig. 9a. The horizontal axis represents "time t," and it begins at 0 (t=0) and moves to the *right as the t increases. The command x shown in Fig. 9 a* means that our goal is for x to jump to its final height k at the instant t=0: in other words, x must jump to its final height in "no time at all" at the instant t= 0. The command x in Fig. 9a is called a *step function* since it resembles a "step" in a stairway. (Because of its sharp rise in "no time at all," it is the most difficult command, or goal, for a machine to follow. Once a machine passes its test of performance for a step-function command, it usually can perform satisfactorily for other commands. For this reason, step-function commands are usually used in the analysis or evaluation of machines.)

For a step-function command x as shown in Fig. 9a, no machines can perform so that the effect or response y reaches its final height in "no time at all." Instead each machine takes a *certain build-up time* t_B to do it, as shown in Fig. 9b or 9c.

Robotosyncrasy 3A: *The build-up time t_B of a machine is inversely proportional to the machine's bandwidth (i.e., mindwidth) W, namely,*

$$t_B = \frac{1}{W}$$

[Unlike all the other robotosyncrasies that can be derived or illustrated with simple mathematics, Robotosyncrasies 3A and 4B require the concept of Fourier Transform for their derivation and are established in Appendix A of Part II.]

According to Robotosyncrasy 1, we know that a robot has a broader bandwidth (or mindwidth) than its crude machine. Assume that the crude machine has a bandwidth $W_1=25,000$ hz and the robot has a bandwidth $W_2=100,000$ hz. We can now compute that the crude machine has a build-up time $t_B=1/W_1$ $=40\times10^{-6}$ secs (or 40 microseconds) and the robot has a build-up-time $t_B=1/W_2=10\times10^{-6}$ secs (or 10 microseconds).

(a)   Command
        Goal-Setting

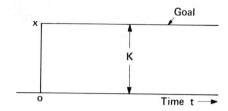

(b)   Effect (Approximate):
        Crude Machine

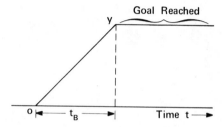

(c)   Effect (Approximate):
        Robot

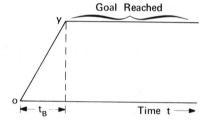

Fig. 9

It is now easy to see that the robot with a larger bandwidth will have a smaller build-up-time t_B than the crude machine, as demonstrated in Figs. 9b and 9c. Robotosyncrasy 3 is therefore obvious.

Robotosyncrasy 4: Using "Opposing Forces" to Achieve Its Goal

The robot uses "opposing forces" to manipulate its effect or response y to approximate the command x.

Suppose that we have the step-function command x as in Fig. 10a. The desired effect or response y is to follow, and be identical to, the command x, as is shown in Fig. 10b; but this is the *ideal* case and cannot be achieved.

In reality, the actual effect y approximates the command x as shown in Fig. 10c or Fig. 11. In Fig. 11, we see positive and negative *overshoots*.

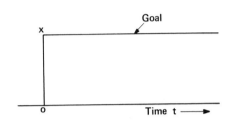

(a)   Command x
       Goal-Setting

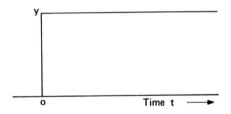

(b)   Desired Effect
       or Response y
       (Ideal Case)

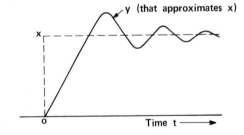

(c)   Actual Effect
       or Response y

Fig. 10

We also note in Fig. 11 that we have positive forces E_a, E_c, . . . (with arrowheads upward) and negative forces E_b, E_d, . . . (with arrowheads downward) to alternately compensate (boost or pull back) the effect y in an attempt to make it equal to the command x.

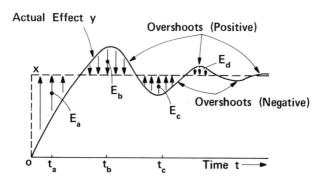

Fig. 11

It is indeed amazing that the robot is able to use "opposing forces" to make its effect y approximate the command x. In other words, these opposing forces actually work together for the robot to achieve its goal.

We shall now also record some basic characteristics of the robot as Robotosyncrasies 4A and 4B:

Robotosyncrasy 4A: *In a robot's attempt to follow a step-function command x (to "achieve a goal"), the "overshoots" in the effect y are inherent phenomena and cannot be avoided.*

Robotosyncrasy 4B: *Only under the "idealized" conditions that the robot has an infinite bandwidth and its gain curve is uniformly flat (like the curve in Fig. 5 but with the bandwidth $W = \infty$), its effect y will be "identical" to the command x (as in Fig. 9c with build-up time $t_B = 0$ and with no overshoots).*

[Like Robotosyncrasy 3A, Robotosyncrasy 4B uses the same derivation for finding the effect y (called e_2 in the derivation) subject to a unit-step command x (called e_1) as in Appendix A of Part II. By letting $\omega_c = \infty$ (equivalent to letting bandwidth $W = \infty$) in the y or e_2 expression, the result is Robotosyncrasy 4B. (We have ignored a small time delay here which is taken into consideration in more detailed descriptions of the work.)]

Robotosyncrasy 5: Visionary and Less Self-Doubt

A robot is more "visionary" and has "less self-doubt" than its crude machine.

As human beings, our mental process is actuated by *signals*

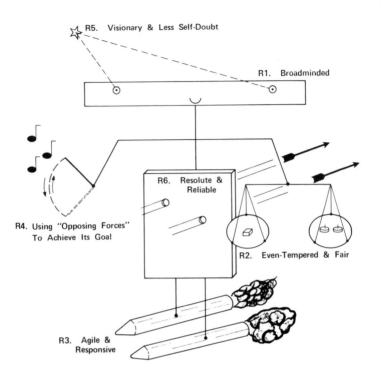

The portrait of a robot

of intelligence and emotion. For robots, the commands given are *signals* and are processed into effects (or responses) which are also *signals*.

A signal is often accompanied by a *noise* (disturbance or doubt) and is corrupted by this noise. If the noise is too large, the signal becomes non-decipherable and meaningless.

Assuming some typical operating conditions and having the "same" internal noise in both a crude machine and its robot, it was found that the robot has far less noise in its effect (or output) than its crude machine.

A typical example shows that for the "same" internal noise, the robot has in its effect (or output) "100 parts of signal and 1 part of noise," while its crude machine has in its effect (or output) "10 parts of signal and 1 part of noise," as is illustrated in Fig. 12.

With much less noise, say, a signal/noise ratio of 100:1, in its output, the robot has an effect (or output) which is much less

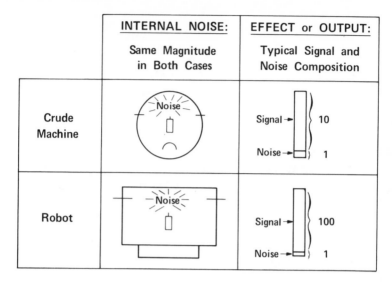

	INTERNAL NOISE: Same Magnitude in Both Cases	EFFECT or OUTPUT: Typical Signal and Noise Composition
Crude Machine	Noise	Signal → ⟩ 10 Noise → ⟩ 1
Robot	Noise	Signal → ⟩ 100 Noise → ⟩ 1

Fig. 12

corrupted by its internal disturbances and self-doubt. A robot is therefore "more visionary" and has "less self-doubt" than its crude machine.

Robotosyncrasy 6: Resolute and Reliable

A robot is much more "resolute" and "reliable" than its crude machine. It maintains a steady gain or vitality factor, even if the robot is "sick" or "wounded" in the sense that many of its components fail to function properly.

The Tains speech in Part II (Chapter 6) shows with some very simple mathematical relationships why this is so. We shall only provide some simple illustrations in Fig. 13 for the end results.

For example, when a crude machine is healthy with all components working, it has gain $G=100$ within its operating frequency range (Figs. 13a1 and 13a2). However, when the machine is "wounded" with several of its components not working, its gain could plunge to $G=20$ (Figs. 13b1 and 13b2). Would it be catastrophic if this machine were a part of the control system in a spaceship? You bet it would!

On the other hand, a robot with all its components working, may have a gain $G^*=50$ within its operating frequency range

(Figs. 13c1 and 13c2). However, when the robot is "wounded" with several of its components not working, its feedback structure and appropriate operating conditions would be able to maintain its gain G*=50 (Figs. 13d1 and 13d2).

Does this remind us of a warrior with deep wounds but still fighting with a full measure of vitality? Resolute. Reliable. And also with a dash of gallantry and chivalry.

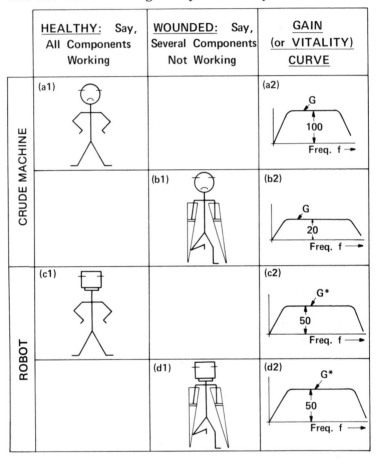

Fig. 13

3. THE CAUSE-AND-EFFECT RELATIONS

A robot is now re-defined to be a machine with a "reflective quality" or a "feedback structure." By *reflectiveness* or *feedback structure* we mean what we have established earlier: It continuously compares what it gets with what it wants. We have also established that a robot has a number of "robotosyncrasies" or "human traits."

We shall find that there are two cause-effect relations for "reflectiveness" (or "feedback structure") and "robotosyncrasies." A third cause-and-effect relation, considered less important for our discussion, is Robotosyncrasy 3A, and will not be discussed here.

A robot deals with a *constituency* of elementary signals in a frequency range *or* more complex signals. A human being deals with different constituencies — signals (of intelligence, thoughts, emotions, etc.), and human groups (e.g., fellow human beings, friends, voters, etc.).

It is believed that the two cause-and-effect relations to be discussed are applicable to both the robot and the human being, although different constituencies will be involved in the interpretation of these cause-and-effect relations.

We shall summarize these two relations from our earlier discussions of the robotosyncrasies in Tables A and B, and then elaborate in the subsequent paragraphs.

Cause-and-Effect Relation No. 1: The "Open Sesame" Approach

If one possesses the human trait of "reflectiveness" (or the existence of a "feedback structure"), *it is "open sesame" to the possession of a host of other human traits or virtues: "broad-*

*mindedness," "even-temper/fairness," "agility/responsiveness,"
"vision/less self-doubt," and "resoluteness/reliability."*
(Table A)

TABLE A

CAUSE	EFFECTS	ASSOCIATED ROBOTO-SYNCRASY
Reflectiveness	*Broadminded*	R1
(or "Feedback Structure")	*Even-Tempered/ Fair*	R2
	Agile/Responsive	R3
	Inherent with Some "Overshoots" (To Achieve A Goal)	R4A
	Visionary/Less Self-Doubt	R5
	Resolute/ Reliable	R6

However, it is interesting to note that, *for all one's "reflectiveness" while attempting to achieve a goal, there will always be "perturbations" or "overshoots"* (or oscillations) *in one's mental process* (Table A). These overshoots are unavoidable.

Now, what is "reflectiveness"? By the robot's rule, it is to continuously compare what you get (the result) with what you want (the goal). The goal differs for different robots or for different persons. For persons, it can be intellectual, spiritual, philosophical, religious, economical, financial, etc.

Nowadays, there are varied activities in transcendental meditation (TM), consciousness raising, positive thinking, and enlightenment by direct intuition through meditation as in Zen Buddhism, and meditation of different forms by different religious sects. By these activities, are we really also "reflecting"?

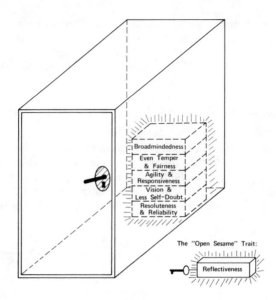

The Fort Knox of virtues as human traits

The Church has taught us always to examine our behaviors as compared to a set of standards. Are we comparing what we do with what we want (or are taught) to do? Again, are we "reflecting"?

The motto of the Rotary Club is the *four way test* for its members. 1) Is it the TRUTH? 2) Is it FAIR to all concerned? 3) Will it build GOOD WILL and BETTER FRIENDSHIP? 4) Will it be BENEFICIAL to all concerned? Again, Rotarians are asking themselves to check their behavior and business conduct against a set of defined goals. Are they actually "reflecting"?

In business and industry, we constantly compare our results (actual sales) with our goals (target sales). "Reflecting" again?

Where does "reflectiveness" lead us? The Cause-and-Effect Relation No. 1 (Table A) tells us that *reflectiveness* as an open sesame trait leads us to these other remarkable traits: We shall become more *broadminded* (in enlarging our constituency of friends, clients, etc.). We shall become *faster* in responding to the demands placed upon us. We shall possess *more vision* and have *less self-doubt*. Even when we have been defeated or wounded, we still have the courage, vitality and chivalry to be *resolute* in pursuing our goals.

For the Cause-and-Effect Relation No. 1, when we possess

"reflectiveness," we can enhance our "broadmindedness" and "even-temper/fairness" to a certain degree, but *not* to perfection in terms of an "infinitely broad" and "uniformly fair" mind. We now introduce Cause-and-Effect Relation No. 2 for this new situation.

Cause-and-Effect Relation No. 2: The "Perfectionist" Approach

If one achieves the ideal situation that one's mind (or constituency) *is infinitely broad* (an infinite bandwidth or mind-width) *and uniformly fair* (equal treatment to all), *then one shall be able to do whatever one wants to do with instant success and no "perturbations" or "overshoots."* (Table B)

TABLE B

CAUSE	EFFECTS	ASSOCIATED ROBOTO-SYNCRASY
Infinite Bandwidth (i.e., Infinite Mindwidth, or Infinite Constituency)	*Able To Do Whatever It Wants to Do With: a) Immediate Result, and b) No "Overshoots"*	R4B
A Uniformly Fair Mind (With Equal Treatment For All)	(Technically, The Effect y is Identical To The Command x With a) Build-up Time $t_B=0$, and b) No "Overshoots")	

In reality one can never extend one's mind to be infinitely broad and uniformly fair. However, one can try to maximize this. And when one does succeed in maximizing it, one shall have a "new freedom" in being able to do *almost* whatever one wants to, with quick (but not instant) realization and very small (but not entirely zero) perturbations or overshoots.

Actually, we have been constantly taught, in the family or in church, to have an *open mind* (broadmindedness?) and to be *fair to all*. If we can faithfully follow these teachings and succeed to maximize these two traits, we shall have a "new freedom" to better control our own destiny.

4. SOCIAL AND POLITICAL IMPLICATIONS

We have associated the robot and its robotosyncrasies with the human being and its human traits. Surprisingly enough, we can also apply the theory of the robot to our social and political institutions. We shall find that

1) Cause-and-Effect Relation No. 1 and
2) Cause-and-Effect Relation No. 2

as discussed above for the human behavior, are also applicable to our social and political institutions. In addition, our social and political institutions also play the role of the robot in terms of what can be called:

3) The Politics of the Impossible, and
4) The Social Progress of Our Society.

Our modern social and political institutions including our various organizations, communities, societies, and governments of different levels, are feedback systems like the robot. In government, our elected officials are our feedback structures. We always compare what we get (the effect or outcome y) with what we want (the command or goal x), and feed the difference (the difference x-y) in terms of our pleasure or displeasure, support or no confidence, or inputs in other forms, on the various issues to our elected officials (the feedback structure). In our other various social institutions, we have various feedback structures sometimes identified with more fashionable terms such as "instruments for accountability."

Therefore, our modern social and political institutions are the same kind of beast as the robot—an entity with a feedback structure. Now, let us see how the cause-and-effect relations (Tables A and B) as discussed earlier for the human behavior are applicable here.

Cause-and-Effect Relation No. 1

If a social or political institution possesses a good "feedback structure," it is "open sesame" to the possession of a host of other institutional attributes: a "large constituency" or large body of support, "equal treatment and protection" for all, "fast realization" in achieving goals, "clarity of the mind" or "less self-doubt" in defining issues and policies, and "abilities to survive crises" even if plagued with internal problems (or being wounded).

Therefore the principal concern for a social or political institution should be how to improve its feedback structure in order to improve its overall efficacy.

But do not expect instant utopia when an institution tries to achieve a new goal. By the Law of Nature, there will be "perturbations" or "overshoots" (Table A). However, if the system is *stable*, these overshoots will die out and this system will ultimately achieve its goal.

(In this writing, we did not discuss "stability." Stable and unstable operations have been discussed in the Tain speech in Chapter 6 of Part II. By design, all robots are stable systems, and the overshoots experienced by them will diminish in magnitude and eventually disappear.)

Cause-and-Effect Relation No. 2

If a social or political institution achieves the "ideal" situation with an infinite constituency ("infinite" in the sense that it has the support of all its constituent members) *and with equal treatment or protection for all* (with no specially favored or oppressed groups or members within its constituency) *then this institution should be able to accomplish what it wants to accomplish, with "immediate result" and no "perturbations" or "overshoots."*

With a large constituent body (say, 100,000 members, or 2 million voters), no social or political institution can command an infinte constituency in that it has the support of every one of its constituent members.

However, it can certainly *attempt* to maximize its constituency and treat all its constituent members equally and attentively, in order to enhance its ability to accomplish what it wants to accomplish with swift (but not immediate) results and with very small (but not zero) perturbations or overshoots.

The Politics of the Impossible

It is only natural that a human being or a social or political institution wishes to achieve its goal meticulously. In the language of the robot, it wants the effect (or result) y (Fig. 10b) to be *identical* to the command (or goal) x (Fig. 10a). But in reality it is an impossible job which will not be permitted by the Law of Nature.

How then are we going to resolve this problem? The robot is apparently able to dabble in this "politics of the impossible" and come up with a solution.

The robot must first realize that it can only achieve to make the effect (or result) y approximate the command (or goal) x, as in Fig. 10c or 11. To do so, the robot employs *opposing forces* (Robotosyncrasy 4) to negotiate (or compensate) the effect y in approximating the command x. The positive and negative overshoots in Fig. 11 are the results of this negotiation (or compensation) carried out by these opposing forces.

Unless the robot has an infinite bandwidth or mindwidth (which is impossible in this physical world), these overshoots are unavoidable (Robotosyncrasies 4A and 4B), and are the natural product of a "negotiative" or "compensatory" process. In other words, although the overshoots are considered by some as "excesses" or "abuses" in either of the two opposite directions, they are actually a *form of adjustment* or a way of life.

Examples in our social and political scenes, where opposing forces are at work with visible overshoots, are numerous:

In our *monetary system* with the Federal Reserve Bank in control, we see periods of "easy money" and "tight money" periods as positive and negative overshoots or cycles to achieve our goals of stimulating business activity and controlling inflation.

On the *political scene*, we note the successive swings of the pendulum between "conservative" and "liberal" eras as adjustments or accommodations for the well being of our nation.

Even in our *foreign policy*, we can readily note the working of opposing forces in alternate dominance with visible positive and negative overshoots, namely, the forces of "isolationism" (or "passivism") and "internationalism" (or "activism"): The Monroe Doctrine (isolationism), World War II (internationalism), Post-War Period (passivism), Vietnam War (inter-

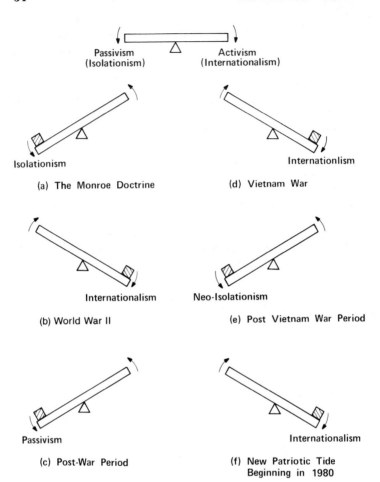

(a) The Monroe Doctrine

(d) Vietnam War

(b) World War II

(e) Post Vietnam War Period

(c) Post-War Period

(f) New Patriotic Tide Beginning in 1980

Two for seesaw or a rhythm for progress? Our foreign policy with the working of opposing forces in dominance

nationalism), Post-Vietnam War Period (neo-isolationism), and the New Patriotic Tide beginning with the New Year of 1980 (internationalism, following the taking of hostages at the U.S. Embassy in Tehran and the invasion of Afghanistan) are examples.

Looking at our *social scene* we note in succession: clean-cut and shaved youth (1940's and 1950's), long-hair, bearded young men and flower girls (1960's and early 1970's), well-groomed and well-dressed generation (now). Are they not the overshoots of adjustment?

Now, consider the *tax scene*: For years, there have been increases and cuts. More recently, the tax burden has become quite excessive. Then the passsage of the Proposition 13 by California voters (1978) as a sign of tax revolt has begun a new cycle of tax reductions. Again, is this a negotiative process by opposing forces with cyclic phenomena?

For the *women's rights*, we can count chronologically: the docile great great great great great granny, the suffrage movement, setbacks, winning of the women's right to vote, women's dominant role at home, feminist movement, popularization of *The Total Woman* (a bestseller) and similar ideas, the Equal Rights Amendment (ERA) ratified in 35 of the required 38 states, anti-ERA and pro-family movements, then what next? Here are many overshoots as the result of opposing forces at work to negotiate the outcome against certain goals.

Considering the movement of *civil rights*, there had already been many upheavals as overshoots of adjustment. For the more recent Bakke case, the U.S. Supreme Court made its decisions (June 1978) that Bakke must be admitted to medical school and that race can be considered as a factor in the deliberation of school admission. The first decision supporting (?) Bakke's "reverse discrimination" charge is an "overshoot" in one direction, and the second decision approving the consideration of race in admission is obviously the next "overshoot" in an opposite direction, perpetuating the overshoots of adjustment.

The Social Progress of Our Society

In many of the above examples, we have listed pairs of "opposing forces" which will work together, guiding the effect (or result) y to approximate the command (or goal) x.

These are opposing forces like *conservative* vs. *liberal, tax increase* vs. *tax cut, isolationism* vs. *internationalism, straight* vs. *hippy*, etc.

Now, which one of each pair of the opposing forces is *right* (correct)? The *conservative*? Or the *liberal*?

Apparently, times have changed. Once upon a time, there was an affirmative answer for each of the questions. The autocratic heads of governments of yesteryear would have no problem of giving you a straight answer. The autocratic bosses of our great industrial firms and businesses of yesteryear would have no hesitancy to give an *unqualified* answer. But those undisputed leaders are a vanishing species. Now we have considerable difficulty to answer: Which of the two opposing forces is right?

Today, there is almost no *absolute standard* for certain measures. As a matter of fact, we are living in a world of "special interests" and "special causes." Except for some involving moral issues, there is no basis for us to judge and declare which of these interests (or causes) is right. Or wrong.

It is amazing that the social progress of our society is made because opposing forces in pairs are working together to guide our effect (or result) y to approximate the command (or goal) x.

It is ironical that some strange bedfellows (as opposing forces) are actually working together for the well being of our society, for example,

> The Selfish vs. The Altruist
> The Greedy vs. The Self-Sacrificing
> The Conservative vs. The Liberal
> The Intelligent vs. The VIP (Very Ignorant Person)
> The Fair-Minded vs. The Obnoxious
> The Reasonable
> Mind vs. The SOB (Stubborn Obstinate Bluff)

And for many special interests or social (or political) causes, the one-track minded OINK's (One Issue Noble Knights), who either know nothing or care for nothing except for a particular issue (as interest or cause), are also contributing in the fashion of

> OINK (for one issue) vs. OINK (for an opposing issue)

toward the well being of our society.

Isn't it interesting (or carzy) that the obnoxious, the greedy, the SOB, the VIP and the OINK along with the rest of us are contributing to our social progress? But the fact is that, in the (robotic) world of negotiative process, they too are essential members of our society for social progress. Former Vice President Hubert Humphrey once said that even the most obnoxious have a right to be heard. Hubert was apparently ahead of his time and was a political philosopher in the Land of the Robot.

It is interesting to note that our democratic (social and political) institutions are alter egos of the robot. Like the robot, their behaviors are prescribed by the two cause-and-effect relations as well as by the negotiative process of using "opposing forces" to achieve their goals.

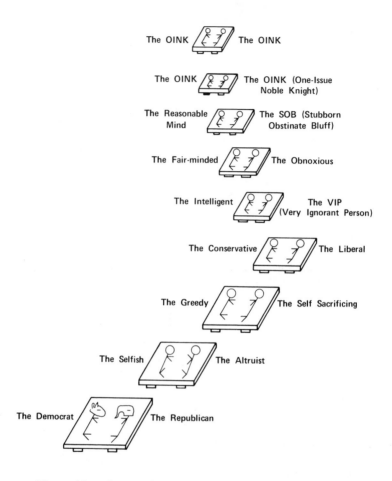

The working of our modern society: strange bedfellows working together for social progress

More Implications

With our discussions on the "Politics of The Impossible" and "The Social Progress of Our Society," we can now make additional observations about the behaviors of human beings and our society.

We may sometimes wonder what are the forces behind a *successful person*. The "robotic logic" now tells us that there are conflicting desires in all of us: the *aggrandizing desires* to make more money, attain higher positions, acquire fame, or achieve more in whatever venture one undertakes (which may be euphorically called "energy and drive," or be referred to in less flattering terms as "greed" or "arrogance"). Or the *sobering desires* to be realistic, offer self-sacrifice, probe and understand situations, realize one's own limitations, and be completely honest with oneself (which may be called "good senses"). A person with all the "greed" but no "good senses" is doomed to failure. And a person with all the "good senses," but no "drive or greed," may not accomplish much. A successful person is perhaps like a well-designed robot with a good "reflective" (i.e., "feedback") mechanism, who has chosen a worthy goal. He reflects, and he continuously compares what he gets with what he wants. And he is able to use his "conflicting desires" in a negotiative process to achieve his goal.

Let's pause for a moment to think about our society or the nation. "Are we actually living in this present society with *impossible demands*"? Just two or three decades ago with segregated buses and segregated public facilities and with governors blocking the integration of schools, it was impossible to imagine a racially integrated society as we have now. The proposed standards for water and air qualities and for automobile emissions, safety and fuel consumption, when first announced during the past decade, were shockers and rebuked by some special groups; but by and large, we are achieving or continuing to achieve what we had thought impossible just a few years ago. On any issue, the original proposal by a particular group of advocates may not necessarily be followed; however, the "negotiative process" by the opposing forces seemed to be able to perform the best possible job (which had once been considered to be impossible). We may feel strange but grateful that the seemingly impossible demands have been responsible for some of our technological, social and political progresses in great strides.

However, the negotiative process no matter how praiseworthy is not a wild free-wheeling process with no restraints. By design, the robot is always a stable system and must behave responsibly. A robot *never overreacts* or behaves intransigently in its negotiative process.

The Conundrum of Hope

Author Lionel Tiger[1] wrote, "optimism, not religion, is the opiate of the people," suggesting that *hope* sustains a person

---

[1] Lionel Tiger, "Optimism, The Biology of Hope," Simon & Shuster, 1979.

(the dying will refuse to die while waiting for a beloved one to arrive) while the lack of hope will decimate him (a baby chimp dies of grief over the death of his beloved mother).

This phenomenon is an age-old conundrum. But why?

The robotic logic explains it very simply: With "hope" one has a goal that stimulates the *dynamics* of one's mind, with positive and negative overshoots to approximate and achieve the goal (e.g., goal x in Fig. 10 or 11). And without hope or goal, one's mind is no longer dynamic. Is this the demise of the mind? Or perhaps the demise of the whole being?

Commentary:

5. THE COMMUNITY LIFE OF ROBOTS

We first compare the human being with the robot, and then the social and political institutions (including governments of different levels) with the robot, for all the robotosyncrasies (or robotosyncratic virtues) as well as for their implications.

How can or does a robot represent these diverse entities? Apparently, there are robots of different orders. A robot with a basic crude machine and a feedback structure shall be called a robot of the first order. Several robots of the first order plus robots of lesser order (crude machines in this instance) with an overall feedback structure shall constitute a robot of the second order, as illustrated in Fig. 14a and symbolically represented in Fig. 14b. Similarly, we may have robots of the third and fourth order as illustrated in Fig. 15 as well as robots of even higher orders.

To show the relationships between the robot and its human, social and political counterparts and their implications, we cite some conversation between Helga, the estranged wife of the robot's fictitious inventor Rob, and her father Frank, from Part II of this book:

"Helga, Rob and I also had many discussions. He told me that by his concept robots themselves can be, and often are, components of another robot of a higher order, and these robots of higher order can be and often are components of another robot of an even higher order. The sophisticated electronic and control systems now in existence often contain robots of very high orders."

Helga listened with great interest. Frank continued: "Think about the implications of this concept. We, human beings, can be the basic robots. With good 'citizen' robots like us as com-

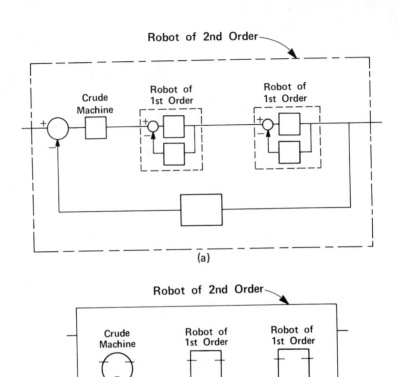

(a)

(b)

Fig. 14

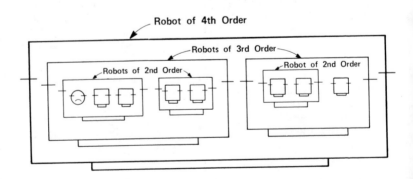

Fig. 15

ponents, our governments and societies of different levels can be robots of higher orders of different levels.

"We design robots to be stable and provide good performance, and a good robot so designed will have all its robotosyncratic virtues. If we set out to design good 'citizen' robots, and then good 'government' or 'society' robots of higher orders, we are more likely to be able to design or produce a good 'world' robot. Who doesn't like to live in a world (call it robot or not) that treats you in a broad-minded, even-tempered, agile, stable, visionary, resolute and reliable manner?"

This is the community life of robots.

Now, let's ask a (tabooed?) question: *Does the robot have a sex life?* In a playful manner, Hawaii has already provided the answer (Part II, Chapter 4).

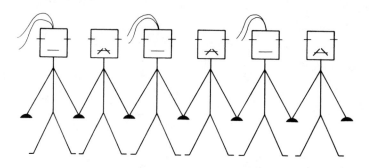

The sex life of the robots: "In a complicated circuit or system, I use NPNs and PNPs (i.e., transistors of the opposite polarities of different sexes in succession. In other words, I 'pair' them in order to achieve sex harmony and make the whole thing work."

Commentary:

6. THE ATTITUDINAL REVOLUTION

Let us consider the brief histories of machines and human beings, and discuss the implications.

Some Recent Developments of Machines

Again, we must deal with the same class of machines that we have dealt with in our earlier discussion. For a given *command*, we wish to produce an *effect* that obeys and follows the command.

Comparing the earlier and more recent machines of this category, we find:

The earlier machines are "open-loop" machines.

The term "open-loop" means that the machine carries out its operation and achieves its aim by using brute force and following dictatorial orders.

The more recent sophisticated machines are "closed-loop" machines.

The term "closed loop" refers to that the machine has a feedback loop (Fig. 2). Machines of higher orders can also have feedback loops in their sub-systems or components which are machines or robots of lower orders (Figs. 14 and 15).

The closed-loop machines or systems first came into existence, perhaps as a "stabilized feedback controller" with a flyball governor connected to the throttle valve of a steam engine for the control of speed, by James Watt over two centuries ago. The principle of (negative) feedback was introduced to vacuum-tube amplifiers making them "stabilized feedback amplifiers" in radio circuits by Harry S. Black in 1934; and a new era or modern closed-loop machines (devices or systems) has begun. During the ensuing four decades, numerous contributions among the engineers, scientists and applied mathematicians have collectively created in an evolutionary process the modern sophisticated *closed-loop machines or systems* which are called under the generic names of their applications: servomechanisms, broadband operational amplifiers, missile guidance systems, process control systems, etc. Closed-loop machines or systems are called *robots* in this book.

The closed-loop machine or robot has a number of operational characteristics. It has a *feedback loop*. It *reflects* in that it always compares what it gets with what it wants. It undertakes a *negotiative process* in its transactions, and uses *opposing forces* to achieve its goal.

Through the transition from the "open-loop machine" to the "closed-loop machine," the machine has undergone through a radical change of its "attitude" or "philosophy of operation." An "Attitudinal Revolution" for machines has begun in the 1930's.

Some Recent Developments in the Human Society

During the recent periods of modern history, we have witnessed some significant changes in our *political institutions*. In recent decades, *business* and *industry* seemed to have changed their management philosophies. And, during recent years, all segments of our *social structure* have been undergoing changes in their modes of operation.

For "political institutions," let's look at the monarchs, republics, and governments of different levels in earlier periods, and compare them with their successor structures after revolutions or evolutionary changes. Inevitably, we note:

The earlier political institutions were often autocratic, dictatorial, centralized and despotic. They usually ruled with *brute force* and *dictatorial orders,* much like an open-loop machine (without feedback).

The successor structures in general take a more democratic form; employ *negotiative* process in their governance through elections and other mechanisms of accountability; and allow *opposing views* (forces) in their processes. They are responsible and sensitive to the needs of the people through their *feedback* mechanisms, much like a closed-loop machine.

For our "business" and "industry" let's have some BEFORE and AFTER snapshots concerning the changes in management philosophies:

The earlier business and industrial organizations were often dominated or controlled by tycoons and moguls who frequently had mixed images. In private live, they were often portrayed as eccentric millionaires (or billionaires). In business, they were euphemistically praised for strong leadership which in stark terms was often meant management by *brute force* and rule with *dictatorial orders* and *iron will,* with very little concern for their employees, clients and the public. Their mode of operation was often not too different from that of an open-loop machine.

More recent modern business and industrial organizations are run by the "organization men" who in gray flannel or pin-stripe vested suits serve as managers at different levels, give-and-take in their *negotiative process* of management, and serve as *feedback* paths to the top management. Information about the market, sales data, public reactions, and impacts of public issues are *fed back* to monitor the conduct of the corporation, much like a closed-loop machine. Also MBO (that is, "management by objectives") which is very much in vogue, is the counterpart of the closed-loop system or robot: It "reflects." And it compares what it "gets" with what it "wants."

Virtually all segments of our "social structure" have been undergoing rapid and profound changes during very recent years, as will be discussed in the next section under the heading of "the Attitudinal Revolution".

It seems that there are momentous trends in our human society involving its political, business and social structures to change from open-loop systems to closed-loop systems much like the machines.

The Attitudinal Revolution

We can take the United States as an example in order to study the changes in the various segments of our social structure.

We have discussed the social and political implications and remarked about several movements in our society during recent years. In addition there are many others, like civil rights, the Vietnam war, women's rights, nuclear installations, taxation, etc.

These movements are really the interplays of opposing forces. Government policies vs. oppositions, the status quo

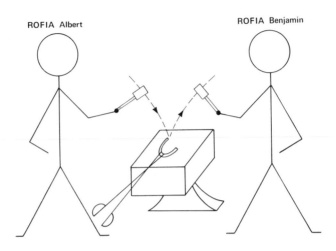

ROFIA Albert   ROFIA Benjamin

ROFIAs in action: seemingly opposing forces in synergetic activity.

force vs. impetus for reform, the charge ahead (for some good cause or technology) vs. caution or restraint, etc. They all involve *Robotic Opposing Forces in Action* (ROFIA) as their mode of principle of operation. And some of these movements have succeeded far beyond any expectation.

Consider the protest movement for the Vietnam War. It was virtually impossible decades ago to imagine that the people (rather than the leaders) could stop a major war, and change our national policy. It offers a beacon of light that major or global wars may be outlawed by the peoples of the world in the years ahead.

Consider the Civil Rights movement. It was equally impossible decades ago to imagine that all the people among the majority and minorities would be now equal before the law. And it offers a gleam of hope that these people may become more and more "equal" in the social and economic arenas in the years ahead.

The United States has actually been experiencing a revolution with enormous impacts: The *Attitudinal Revolution of the Robotic Kind* (ARRK) permitting and promoting the negotiative process using opposing forces. This revolution is already having an impact on the world through the vanguard of "human rights."

We see that, in many instances, the majority interest can prevail in the policies and governance in the governments of different levels, rather than be dictated by a few strong leaders. The minority interest can be also protected through the principle of ROFIA.

This principle makes us re-examine all our traditions and practices. Consequently, it can minimize ignorance. (However, ignorance is a special gift of the human race and cannot be totally banned. The world would certainly be much less interesting without some of the tantalizing effects of ignorance.) It can tear down traditions that are forms of unreasonable restrictions, prejudices, and oppression. It can help create a society with sensitivity, consciousness and awareness, which are robotosyncratic virtues in another form. It can send tyrannical governments into oblivion.

The United States has been criticized in recent years for its

ROFIA in action: helps create a society with sensitivity, consciousness and awareness.

lack of law and order, for being a weak leader, and for not taking a hardline approach in diplomacy and military matters. These criticisms are not well based. *The United States is actually leading the world in a new revolution, the ARRK.* The government is becoming *less* of an open-loop machine that governs by brute force or dictatorial order or insensitive edict. Through the foresight of our ancestors who wrote the Constitution, the United States government has never been an entirely open-loop machine. Instead, it is becoming *more* of a closed-loop machine or intellectual robot that "reflects" about its own performance and has all the robotic virtues discussed earlier, namely, how to be broad minded and fair, how to be responsive without overreacting, how to be visionary and deny self-doubt, how to be resolute and reliable, and how to allow negotiative process using opposing forces.

For the United States the new revolution may make her a democracy in the true sense. A new age has dawned on us. The greening of America? Or merely the blushing of a new springtime?

With the U.S. leadership, the ARRK may someday sweep the world. More democracy? More freedom? More of an equalitarian society? Major wars may be banned for ever? Could this be the beginning of a utopian society? A "one world" as envisaged by Wendell Wilkie? Or a transitory phenomenon? Or is this merely a mirage with an ephemeral splendor?

It is amazing and interesting to note that an Attitudinal Revolution began among the machines in the 1930's. Today, we are carrying their banner and extending this revolution in the human world.

The Safeguard and Limitation in the Attitudinal Revolutions

Many of us are concerned whether the protest movements are going too far and too fast. Are the Robotic Opposing Forces in Action (ROFIA) becoming a menace to our civilization? And is our ARRK negotiating the dangerous waters of change and coming down the rapids out of control and in the danger of capsizing—along with the crumbling of the human society?

Fortunately, there is a *safeguard* in the robot's rules book to *limit* the speed of change. According to Robotosyncrasy 3A,

The build-up time t_B, as required to achieve one's goal, is inversely proportional to the bandwidth W (namely, $t_B = 1/W$).

Since what takes *more* time to achieve has a *slower* speed, we therefore conclude that

> *The larger the bandwidth W, the faster the allowed speed of change.*

It is interesting to note that unless an entity (say, a person, a society or a government) has a sufficiently large bandwidth (say, a large mindwidth, or a large "constituency" or a large body of support), its allowed speed of change is rather limited. On the other hand *if its bandwidth can be broadened, its allowed speed of change is proportionally increased.* (modified Robotosyncrasy 3B)

The fall in early 1979 of Shah Mohammed Reza Pahlavi from the Peacock Throne of Iran was due to many reasons. However, it is obvious that the Shah had violated the above modified Robotosyncrasy 3B. He had wanted to move Iran into an ultra-modern country with too great a speed (or too short a build-up time) without cultivating a sufficiently large bandwidth or body of support. According to the robotic law, the government would become an *unstable* system, oscillate violently and eventually break down. (In contrast, all robots among machines are by design *stable* systems; if one of the preliminary designs is found to be unstable, it is always rejected or redesigned.)

It is comforting to know that an attitudinal revolution is ready only when there is a sufficiently large bandwidth or body of support.

The Jekyll and Hyde of the Robot

For a machine to be a robot, it needs to be "reflective." The results of this "reflection" (or feedback feature) are the robotosyncrasies discussed earlier.

Like *Dr. Jekyll and Mr. Hyde*, a robot also has a dual personality (however, without a "vicious" element): the *soft* side and the *hard* side.

The "soft" side of the robot consists of these robotosyncratic features: *broadminded, even-tempered, fair, agile* and *responsive*; while the "hard" side of the robot has these features: *ROFIA* (robotic opposing forces in action), *visionary, least of self-doubt, resolute,* and *reliable*.

We have discussed the application of the "robotic laws' to different "levels" or "orders" of entities, e.g., the *person, com-*

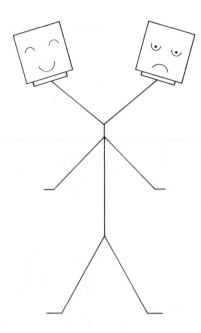

The dual personality of the robot: the soft side and the hard side.

munity, society, governments of different levels, the *nation* and the *world*. Some of these applications are novel and have resulted in discussions given in earlier sections.

Have some of the robotosyncratic features been *misapplied* or *misdirected*? Let's now examine the situation, in particular, on the level of a "person":

Once upon a time (in the pre-ARRK days) a typical "person" had these characteristics:

> His (or her) activities involved primarily *personal conduct* with relationship with others; he (or she) was "reflective" and had most of the robotosyncratic virtues, particularly those on the *soft* side. (These were the old virtues as the moral fibers of our society.)

Later, in the ARRK (Attitudinal Revolution of the Robotic Kind), this change occurred:

> He (or she) became also interested as a *participant in public issues* (as a proponent, opponent or activist). He (or

she) became fascinated by *ROFIA* (robotic opposing forces in actions) and the other *hard* features of the robot.

"Participation in public issues" and "ROFIA" made good chemistry, precipitating the ARRK with all its perquisites.

However, his (or her) penchant for *ROFIA* and other *hard* robotosyncratic features was carried over to his (or her) *personal conduct*, disregarding the soft "old virtues" and shredding the "moral fibers" of our society. He (or she) became inconsiderate, arbitrary, unfair, selfish, unfriendly, and belligerent to his fellow human beings (in the family, school, working place, community and other public activities) but had a strong *misdirected* or *misapplied* conviction that he (or she) was a champion and revolutionary hero of ROFIA.

The misdirected conviction of our hero was perhaps at least partly responsible for many of our societal problems.

The robot (as a machine) has a dual personality and plays both Dr. Jekyll's and Mr. Hyde's roles (without the viciousness, actually playing Dr. Hard and Mr. Soft with their accompanying characteristics) well. We, the human imitators, fail to understand the uniqueness of each role:

1) For "personal conduct" with relationships with others, play *soft,* and

2) For "self-discipline" and "participation in public issues," play *hard.*

Nowadays, there is no alternative but to play "hard" on ourselves for *self-discipline,* as the robotic laws do not allow us to apply double standards in playing spoiled brat in private life and hero on public issues.

The robot has again taught us how to improve ourselves for a renaissance of the whole man (with "soft" compassion and "hard" self-discipline) as well as of our family, community, and society at large.

7. THE MODERN ETHICS

Traditional Morality

Traditional morality has many stringent standards of conduct, and they are taught and practiced at home, in the church and in some schools.

Among these standards are broadmindedness, even temper and fairness, agility and responsiveness, vision and lack of self-doubt, resoluteness and reliability. It is interesting that traditional morality consists of all robotosyncrasies *except* for Robotosyncrasy 4 that allows ROFIA (robotic opposing forces in action), or, using "opposing forces" to achieve the goal.

Despite all the varied standards, there are two overriding idealized criteria for traditional morality:

1) *Even if you are 99% right, try to be also right for the remaining 1%.*
2) *Behave like a paragon for others to emulate.*

We have heard these criteria over and over for generations from our elders, and have respected them until recently.

In the context of traditional morality, we are taught to achieve various high standards of conduct. However, we are not taught *how to* achieve these standards.

Fortunately, the robot *has* taught us. Recall the "Cause-and-Effect Relation No. 1" in Chapter 3: we know that we must first develop *reflectiveness* and that it is "open sesame" to the possession of a host of other human virtues or high standards of conduct.

It is not surprising that in the various consciousness-raising activities, in church, in social circles, and in business and industry, we are often taught to *reflect*. It is an essential tool for us to reach the summit of traditional morality.

Modern Ethics

Modern ethics are different. They do not disavow the various stringent standards of conduct. However, they do scrap the two overriding criteria for traditional morality and replace them with two new criteria of their own:

1) *If you are 1% right, champion your cause on this 1%.* Or if you are right (or believe that you're right) on one issue, champion this issue.

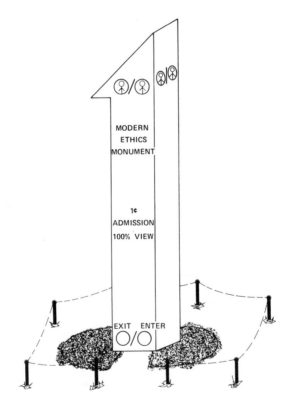

Modern ethics motto: "I know 1% of the truth. But truth is truth, and dammit, you'd better listen to my truth."

2) *Damn the concept of behaving like a paragon for others to emulate. I'll do whatever I please and it's none of anyone's business.*

Apparently, Robotosyncrasy 4 has now been discovered, allowing ROFIA and using "opposing forces" to achieve the goal.

The Age of Collective Wisdom

We now have the "one percenters" or "one-issuists" or OINKS (One Issue Noble Knights). Yes, if you are 1% right (i.e., "right" for 1 issue out of a 100 issues), you are entitled to champion your cause. If the establishment is 99% right but wrong in 1% (and that's the 1% you are right about), challenge it with all that you have, in vehement terms if you wish.

But, what is the justification for these "one percenters"?

Remember that society is becoming more egalitarian and there is now almost no absolute standard (right or wrong) for many measures. How then are we going to set policy or charter courses of action for our government or other establishments? We must use the principle of ROFIA, i.e., robotic opposing force(s) in action. And the "one percenter" is a ROFIA, exercising his or her prerogative to contribute to our society. It is now the *age of collective participation* or *collective wisdom.* As mentioned in Chapter 4, we may feel strange but grateful that the many seemingly impossible demands by these "one percenters" have been responsible for some of our recent strides in technological, social and political progress.

In reality, there are not too many pure-bred traditional moralists or modern-ethic one percenters. Many of us are becoming hybrid types – the mulattoes of these two breeds. Aren't we fortunate to have some one-percenter streaks (OPS) that allow us to champion like a ROFIA at the opportune moment, and then subject ourselves to the weight of the traditional morality at other times. The OPS is a new phenomenon and is necessary to us, as a catalyst, for progress.

While in praise of ROFIA, we must be aware of MAFIA (Mal-Adjusted Force(s) in Action). By design, the robot is always a stable system and must behave responsibly. This means that a ROFIA never overreacts or behaves intransigently in its negotiative process, thus contributing to the overall system stability and behavior. On the other hand, a MAFIA *overreacts* and *behaves intransigently,* causing the

system to be unstable and break down. If you have MAFIAs
out of control, be prepared to pick up the pieces of a ruined
society.

Commentary:

8. THE TEN COMMANDMENTS

The author has not been able to discover a biblical origin of the robot. However, on the basis of our earlier discussion, it is obvious that the robot has its own ten commandments. Inasmuch as we have applied the "robotic laws" to different levels or orders of entities, the *person, community, society, governments* of different levels, the *nation* and the *world*, these commandments are directly applicable to the conducts of all these entities. These ten commandments are:

1) Thou shall be *reflective* in probing all thy conducts, and compare what thou "geteth" with what they "wanteth."

The above commandment is often considered the "master" or "open sesame" commandment. The fulfillment of this commandment will automatically fulfill the other commandments to a certain degree—but not completely. For this reason, the other commandments should still be strictly exercised and followed for their complete fulfillment.

2) Thou shalt be *broadminded*.

3) Thou shalt be *even-tempered* and *fair*.

4) Thou shalt be *agile* and *responsive*.

5) Thou shalt allow some *overshoots* (or "oscillations" or "perturbations") in thy emotions and efforts.

Commandments 2 through 5 may be considered the "soft" package among the commandments. Although all commandments are important to all conducts of all entities, the soft commandments are particularly important to the conduct of a person and are the old traditional virtues which are the moral fibers of our society.

6) Thou shalt allow and encourage *ROFIA* (or "robotic opposing forces in action") in all thy endeavors.

7) Thou shalt not *overreact* and shalt *not* be *intransigent*.

When you *overreact* with a large "overshoot," it will be followed by large "overshoot" in the opposite direction (Fig. 11). The *larger* your overshoot is, the *larger* the succeeding one in the opposite direction is. What have you accomplished? You have just added strength to your opponent. Don't be mislead to think that you are the only wise guy in the world and all others are fools, and that when you knock them hard they will not answer in kind. And when you *overreact* and are *intransigent*, the successive overshoots will grow in size and you have an *unstable system*, which will break down.

Now, what is the remedy? Follow the other commandments, be compassionate, be flexible, negotiate peacefully, and coexist with others.

8) Thou shalt have *vision* and the *least of self-doubt*.

9) Thou shalt be *resolute* and *reliable*.

Commandments 6 through 9 may be considered the "hard" package, which is important for self-discipline.

10) Thou shalt possess the *judgement* when to play soft, when to play hard, and when to play both.

Commandment 10 tells us that we have a choice for our behavior. On certain matters (particularly, those concerning public issues), it is even all right to play the SOB (Stubborn Obstinate Bluff) or to behave like a spoiled brat. However, it is important to play only Brat First Class (BFC) with some high principles: a) Let go your worst temper tantrums only in some confined public-issue domains and don't let it spill over to other matters. b) Don't ever carry your stubbornness to the ultimate state of intransigency. c) Remember your exalted rank as BFC and don't behave like just any rank-and-file spoiled brat.

Commentary:

9. THE VARYING ROLE OF THE ROBOT

What has been said so far is a *true* story. Together with the novel (Part II), it may be considered as a biography for the robot. However, the robot has been cloned. There are now billions of robots in existence, all with a set of robosyncratic virtues as their genetic features. They are the workhorses in electronics, communication, computer and aerospace systems. It is no exaggeration to say that, if all these robots were taken out of their present stations, the world would come to a grinding halt.

Robot, the Sage, has taught us how to cultivate a particular human trait — reflectiveness, — and as a consequence we shall possess a host of other human traits. He has also taught us how to cultivate two human traits to perfection (an idealized but not realizable condition), and we shall then be able to do whatever we want with instant success and no perturbations.

Robot, the Prophet, has taught us that there is nothing absolutely right or wrong, and that we can never in reality reach our goal instantly and without perturbations (or overshoots). He prophesies that we should use "opposing forces" to negotiate our way to reach our goal as a wave of the future.

Robot, the Mentor, has taught us when to play soft, when to play hard, and when to play both.

Robot, the Revolutionary, has taught us how to change our society and how to change human destiny.

Technology has changed our lives enormously through its wonderful products. Now, through the robot, it is encroaching upon the turfs of religion, morality and philosophy to teach us how to behave, how to become a whole man or woman, how to bring about a renaissance of the family, community, and society at large, and how to guide our own destiny.

A blessing? Or a threat? We don't have to answer these questions. The robot has already told us that nothing is absolutely right.

QUIZ/GAME

The following quiz is for you, the reader, to determine your knowledge of the intellectual robot.

Please choose "true" or "false" for each of the statements on the Quiz sheet. Look up the keys to this quiz, and add up your score. Then fill our your CERTIFICATE OF RECOGNITION. Good luck!

The Quiz

	Statement	*Answer*
1.	SOBs (Stubborn Obstinate Bluffs) are *essential* to our social progress.	T F
2.	When you have a *broader* mind, you are *faster* in responding to a demand or goal that you have set to achieve.	T F
3.	You must have *both* "greed" and "sobering desires" to be successful.	T F
4.	To set a *seemingly impossible* goal is important for reaching a *higher* level of achievement.	T F
5.	MBO, i.e., management by objectives, like a missile guidance system, is a *closed-loop* system.	T F
6.	If you *"reflect"* on your own performance you shall have the *least* of self-doubt.	T F
7.	VIPs (very ignorant persons) will question everything thus *preventing* hasty actions and mistakes by the impulsive leaders, and are *therefore* VIPs (very important persons).	T F

8. The 3-Mile Island nuclear accident was an *impor-*   T F
   tant contributor to the development of nuclear
   energy; when the energy crisis reaches a critical
   state and the nuclear energy is again rediscovered
   by the people, we shall have *safer* nuclear reactors
   to serve us.

9. The principle of ROFIA (robotic opposing forces in   T F
   action) is a *mechanism* of checks and balances, and
   is *important* to the governance of our nation.

10. To achieve an *infinitely broad* mind and to be *uni-*   T F
    formly fair to all is not just a religious exercise. By
    the Law of Nature, it enables one to do *whatever
    one wants* to do with instant success and no pertur-
    bations (disturbances).

11. The conservative and the liberal are *strange bed-*   T F
    fellows, working together for the well being of our
    society.

12. The Law of Nature dictates that in our attempt to   T F
    achieve a goal, *perturbations* (disturbances) are in-
    herent phenomena and cannot be avoided.

13. One of the most important *secrets* of life is reflec-   T F
    tiveness, and *reflectivensss*, once developed, could
    spawn a host of desirable human traits.

14. MAFIA (Mal-Adjusted Force(s) in Action) is a   T F
    spoiler, and it could *ruin* our society through its
    intransigence.

15. In modern ethics, you are *entitled* to champion   T F
    your cause if you are 1% right. The establishment
    should be *damned* even if it is 99% right (but is
    wrong for the same 1% that you are right for). Sock
    it hard to them!

16. A simple algebraic equation $G* = G/(1+GH)$ that   T F
    supports the underlying principles of feedback
    amplifiers, servomechanisms, process control
    systems, and missile guidance systems, also *ex-
    plains* many of our social phenomena which had
    been often considered inexplicable, and *teaches* us
    how to behave.

17. The (intellectual) robot knows when to play *soft,*   T F
    when to play *hard,* and when to play *both.*

18. In the robotic world, *nothing* is absolutely right.   T F

19. The principle of ROFIA will decide most of the   T F
critical issues facing us, allowing *opposing forces*
to be alternately in dominance guiding us to
achieve our collective goals.

20. The robot has taught us how to improve ourselves   T F
for a *renaissance* of the whole man with soft compassion and hard self-discipline.

21. The robotic logic warns us: *Don't overreact and be*   T F
intransigent in perpetuating overshoots or excesses in opposite directions (like Romeo's and
Juliet's families). *Accommodation* is the best
policy.

22. If one is *intransigent* and hits one's opponent hard   T F
(as with a large "overshoot"), the opponent may be
induced to a *stronger resolve* (a larger "overshoot"
in the opposite direction) in reaction. Think hard of
some example of international politics where the
intransigence of one government *rallies* another
nation to unity and patriotism.

23. The principle of ROFIA (robotic opposing forces in   T F
action) has made it possible for the people (not
their leaders) to *stop* the Vietnam War. An omen
for the banning of major wars around the world in
the distant future.

24. An Administration with all "yes" men *cannot*   T F
aspire to and fulfill lofty goals, as its ROFIA are
weaklings and have no strength of their own to
help achieve the goal through negotiative process.

25. The United States is leading the world in a *new*   T F
revolution, the Attitudinal Revolution of the
Robotic Kind (ARRK), using the principle of
ROFIA to make the government and society more
responsive to the needs of the people.

26. The technology of the robot (i.e., the feedback   T F
system) is *ubiquitous* and *pervasive*. There are
billions of these robots as workhorses in electronics, communications, computer and aerospace
systems. If all these robots are taken out of their
present stations, the world would come to a grinding halt.

27. The landslide election of Ronald Reagan as Presi-   T F
    dent in November 1980 was *no accident*; appar-
    ently a change, as predicted by the robotic law,
    was overdue.

28. Hope is known to sustain a person while the lack of   T F
    it will decimate him. Why? This age-old *conun-
    drum* can be readily explained with robotic logic:
    With "hope" one has a goal (e.g., goal x in Fig. 10 or
    11) that stimulates the *dynamics* of the mind. And
    without hope or goal, one's mind is no longer
    dynamic; it dies.

29. The U.S. Supreme Court decision (June 1978) that   T F
    Bakke must be admitted to medical school and
    that race can be considered as a factor in the
    deliberation of school admission, is really allowing
    overshoots of adjustment in opposite directions for
    accommodation.

30. In view of the fact that there were thousands of   T F
    robots in the various electronic and control
    systems of *Apollo 11* and the mission was a com-
    plete success, it is therefore appropriate to
    paraphrase Neil Armstrong's well known remark
    during his moon walk to pay our tribute to the
    robot: "One small step for the robot, a giant leap
    for mankind."

Academic Honors and Requirements/Keys to the Quiz

Academic Honors	Requirement (Correct Answers)
Robot of Philosophy (PhR or ROPH as "ruff")	27-30
Master of Robotic Arts (MRA or RAM)	23-26
Bachelor of Robotic Arts (BRA or RAB)	19-22
Robotic Associate Degree (RAD)	13-18
Robotic Drop-Out (RODO)	7-12
Robotic Illiterate Being (RIB)	0-6

Your score is _____

Key to the quiz: All 30 statements are "true."

Your academic honor is _____

CERTIFICATE
OF
RECOGNITION

This certifies that _____
has demonstrated his or her knowledge in robotics
through self-graded competitive examination and is
now accorded the academic recognition as:

_____

with all the honors, rights and privileges thereof
apertaining.

THE INTELLECTUAL ROBOT

Date _____

APPENDIX I-A

In this Appendix we shall report, in a narrative form and with a minimum of scientific terminology, the "robotosyncrasies" or "human traits" of the (intellectual) robot.

(These robotosyncrasies are mathematically derived in Chapter 6 plus Appendices A and B of PART II of the book. And they are given merely as "scientific facts" with no mathematics, but with scientific terminology, in Chapter 2 of PART I.

We are dealing with a special class of machines that when we give a *command* (or *input*), we would like to have the machine's *response* (or *effect*) to obey and follow the command (or input). In other words, we want the "effect" to duplicate the "command."

For example, we may provide an "input" to an amplifier, say, in our hi-fi set, and wish to produce an "output" or "response" which duplicates exactly the input in order to preserve the quality of high fidelity. *Amplifiers* are used not only in hi-fi sets; they are used everywhere, say, in radios, televisions, radars, computers, electronic systems, aerospace systems, weapon systems, medical equipment, etc. Other machines belonging to the same class are: *servomechanisms, closed-loop systems, feedback control systems, process control systems, missile guidance systems,* etc.

The Robot

A *"crude machine"* designed for the above purpose is schematically shown by a G-box in Fig. 1 (Chapter 2), where the command is denoted by the symbol x and the response, by y.

A *robot* (or *intellectual robot*) is a modified form of a crude machine. It continuously compares what it gets (the effect y) with what it wants (the command x) through a feedback look, and feeds their difference to the crude machine G, as shown in Fig. 2*a*. A more elegant robot is shown in Fig. 2*b*.

For both "crude machines" and "robots," their commands are *signals* (while their responses are physical quantities, e.g., position, speed, etc., with the shapes or waveforms as "signals").

Through a simple algebraic equation $G* = G/(1+GH)$ for the robot in Fig. 2*b*, we can readily derive a number of "robotosyncrasies" or "human traits" of the robot.

Robotosyncrasy 1: Broadminded

A robot is much more "broad-minded" than its crude machine.

A robot is more broadminded in that it accommodates a larger number of its "constituents" (i.e., elementary signals). We say that it has a broader "mindwidth."

Robotsyncrasies 2: Even-Tempered and Fair

A robot is more "even-tempered" than its crude machine.

In other words, a robot is more "fair" to its constituents (i.e., elementary signals) in that it treats them more equally.

Robotosyncrasy 3: Agile and Responsive

A robot is much more "agile" (faster) in responding to a command than its crude machine.

In other words, for a given command (goal setting), the robot reaches its goal "faster" than its crude machine. It is more "responsive" to the demand or goal that it has been asked to achieve.

Robotosyncrasy 3A: The time for a machine to achieve its goal, designated as t_B here, is equal to

$$t_B = \frac{1}{W}$$

where W is the machine's mindwidth.

Robotsyncrasy 4: Using "Opposing Forces" to Achieve Its Goal

The robot uses "opposing forces" to manipulate its effect y to approximate the command x.

In reality, if the command (or goal setting) is x in Fig. 11 (Chapter 2), the effect cannot duplicate the command exactly. Instead, the effect y will "approximate" the command x, as shown in Fig. 11. Note that there are positive and negative "overshoots."

Note that there are "opposing forces" guiding the approximation process:

1. When y (namely, its height) is *smaller* than x, there are *positive* forces E_a to pull it up.
2. When y is apparently "overpulled" in the upward direction and is now *greater* than x, there are *negative* forces E_b to pull it down.
3. When y is apparently "overpulled" in the downward direction and is now *smaller* than x, there are *positive* forces E_c to pull it up again.
4. When y is apparently "overpulled" in the upward direction and is now *greater* than x, there are *negative* forces E_d to pull it down again.
5. This process *repeats* itself with *smaller and smaller overshoots*. The response y (namely, its height) eventually *equals* the command x (namely, its height).

Here we note that the robot uses "opposing forces" to achieve its goal.

Robotosyncrasy 4A: In a robot's attempt to achieve a goal, the "overshoots" in its response are inherent phenomena and cannot be avoided.

Robotosyncrasy 4B: Only under the "idealized" conditions (namely, mathematical conditions, which cannot happen in the real world) *that 1) the robot's mind is infinitely broad* (in that the robot can accommodate all its infinitely many constituents – elementary signals) *and 2) the robot's mind is uniformly fair to all its infinitely many constituents, the effect y will duplicate the command x exactly.*

Robotosyncrasy 5: Visionary and Less Self-Doubt

A Robot is more "visionary" and has "less self-doubt" than its crude machine.

In other words, a robot will have much less "noise" (disturbance or doubt) in its response than its crude machine.

Robotosyncrasy 6: Resolute and Reliable

A robot is much more "resolute" and "reliable" than its crude machine.

Amazingly, even when a robot is "sick" or "wounded" (for example, many of its components are burnt out), the robot can still function properly. This is not the case for a crude machine.

Part II

The Robotosyncrasies

Introduction

This is perhaps one of the oddest books ever written.

1) It is the author's intention to bring to the reader one of the great scientific creations, which has an enormous impact upon our life ("Without it the world would come to a grinding halt!") and which actually has human intelligence. However, this amazing fact is neither known to the general public nor obviously recognized even by the engineers/scientists. This message, given as the technical keynote speech complete with discussion and figures in Chapter Six, may be used as a "reading assignment" for high school or college students or read by the general reader to learn and appreciate the workings of our technological society. Sections of an advanced engineering book are included as Appendices A and B for providing authenticity and reference. Because of its human intelligence, we shall dub this creation as a "robot."

2) The balance of this book is a fiction, a love story of an engineer/scientist who had created a robot with human intelligence. And it is also a satire about the space efforts and the fate of the aerospace engineer/scientist with a scenario landscaped with the social flora of our time. The reader may omit the technical portion in Chapter Six in reading this book.

Author's Remarks About This Scientific Creation, the "Robot"

This great scientific creation is a relatively simple system model. Using this model and with the aid of simple algebra (as studied in the seventh or eighth grade), it is *mathematically proved* (rather than merely given) that this creation has many human qualities, e.g., as being *broad-minded, even-tempered, agile, stable, visionary, resolute* and *reliable*. And there are also many other implications. For example, if the creation, dubbed as a robot, receives a certain "command," nature does not allow the effect or response to be entirely identical to the command and instead the "effect" only approximates the command and has positive and negative overshoots. Doesn't this sound rather familiar in terms of the successive swings of the social pendulum in the process of social evolution?

This scientific creation, now called a robot because of its many human qualities, is a true story and is *not* a science fiction.

The robot is actually being used as a workhorse, numbered in billions, in electronics, communications, computer and aerospace systems. It is no exaggeration to say that, if all these robots were taken out of their present stations, the world would come to a grinding halt.

The robot is really the creation of numerous contributors among the engineers, scientists and applied mathematicians in an evolutionary process throughout the years. Partly because it is the cumulative effort of a large number of contributors and partly because the engineers and applied scientists are quite "sloppy" in their documentation, almost no engineering textbooks give references and credit to the creators of the robot. And the robots are designed and used extensively, assuming that the design information is a "common body of knowledge" to be used freely without having to give any credit lines.

Even the contributors to the creation of the robot have actually never used the term "robot"; instead they call it with a variety of names, e.g., vacuum-tube feedback circuit (or amplifier), transistor feedback circuit (or amplifier), feedback system, feedback structure, automatic control system, closed-loop control system, servomechanism, fire control system, position control system, speed control system, etc. And these contributors have never thought of it in terms of human qualities; instead they think of it in terms of technical properties for optimal operations.

It was the author's interpretation through the words of the novel's protagonist that this created entity, now called a "robot," has human qualities. And these human qualities are actually mathematically derived from the robot's simple system model.

The protagonist of this novel, credited as the creator of the robot, is merely a figment of the author's imagination. This book is dedicated to the unsung heroes of the great creation.

One

Once a year in late May, the Institute of Engineers and Applied Scientists (IEAS) holds its international convention in New York City. With a million members distributed throughout the world, the IEAS is the largest professional society of its kind.

Tonight was the banquet night of this year's convention. The grand ballroom of the headquarters hotel was literally full of people sitting around numbered round tables. Theirs were all happy faces marking a very auspicious occasion, for this was the 100th anniversary of the giant IEAS.

The head table on a raised platform was a long one seating twenty-two people who had reasons to be there. Two speaker's rostrums had been installed, one at the center of the head table for speakers and the other at the end of the head table on the right for Dr. John Ryder, vice-president of the IEAS, serving as toastmaster tonight.

Dr. Frank Baker, president of the IEAS, sat on the immediate right of the center rostrum. Jack "Rob" Tains, who sat next to Dr. Baker, was understandably a little nervous for all the attention that had concentrated on him throughout the convention. Only half an hour ago at the IEAS president's reception in the penthouse of the hotel, he was photographed by request with the various dignitaries for national wire as well as for local publicity of these dignitaries in their hometowns. Also, at this reception, he met Helga McGee in person. He had been intrigued by her mellifluous voice and her direct approach when she called this morning over the phone for an interview.

"Dr. Tains, I am Helga McGee of the *New York Tribune*. Are you the 'robot' authority who is quoted to 'use the robot principle to dictate human destiny'?"

"Ms. McGee, I am merely an engineer, or a scientist as some people call me, trying to apply some scientific principles dictated by nature to technology serving mankind——"

Without waiting for a full response, she injected, "I believe that we have a great deal to talk about."

"I guess so," Rob replied.

"How about ten tomorrow morning? I'll come to your Manhattan office."

"Why not?"

So he was to be interviewed by Helga McGee.

Beef bouillon, prime ribs, and rum cheesecake, well prepared, and which lived up to the reputation of the hotel. The program now began in accordance with the provisions spelled out in a thirty-page "banquet program."

After a brief welcoming speech by President Baker, Vice-President Ryder as toastmaster began to recognize the newly elected fellows of the IEAS who were present at the banquet. Selected one out of 5,000 members (or 200 out of 1,000,000 members) for distinguished achievements, the "fellow" grade was the highest distinction that the IEAS could bestow upon its own members. All these new fellows and their spouses sat at the round tables close to the head table. As Dr. Ryder called their names individually, they stood up to be recognized.

When the roster was completed and a round of applause subsided, a representative of the new fellows who sat at the head table, Dr. Irene Peden, walked to the rostrum and responded in behalf of the group for this recognition. In obvious reference to Dr. Ryder's introduction of her as "the most beautiful of the new fellows," she blushed and said, "Beautiful or not, we are grateful for this beautiful recognition of our past contributions and shall strive to continue our contributions for the welfare of mankind."

The program then continued through four major awards with presentations, acceptance speeches, and appropriate panoplies.

Now, it came to the main events of the evening. The theme of this year's convention was "a century of progress." There were three keynote speeches on the same theme and crowning the presentation of the Centennial Medal of Honor award.

The Medal of Honor is the highest award of the Institute of Engineers and Applied Scientists, and is presented only when a candidate is identified as having made a significant contribution which forms a clearly exceptional addition to the science and technology. Ninety-nine persons have been awarded the Medal of Honor in the past ninety-nine years. A complete list of these ninety-nine recipients was published in the banquet program.

Among these ninety-nine awards, three were identified as the Quarter-of-the-Century Medal of Honor awards, recognizing the most significant contributions during the twenty-five-year periods and marking the twenty-fifth, fiftieth and seventy-fifth anniversaries of the IEAS.

Known by his friends and associates as Rob (pronounced as "robe"), Dr. Tains sat somewhat uncomfortably as Dr. Ryder introduced this portion of the program.

Rob thumbed through the program, and read the list of the earlier ninety-nine recipients of the Medal of Honor awards. He had the ambition to become an engineer/scientist ever since he was five. He was the admirer of an array of great inventors—the inventors of adding machine, airplane, automobile, clock, Diesel engine, steam engine, gyroscope, incandescent light, locomotive, microphone, movie, photography, printing press, radar, radio, vulcanized rubber, spectroscope, steamboat, steel, submarine, telegraph, telephone, telescope, transformer, turbine, x-ray. To his amazement and awe, more than half of these inventors received the IEAS Medal of Honor awards.

As the current recipient, Rob mused, "Am I in their ranks now?"

As he was musing, Dr. Ryder intoned, "On our twenty-fifth anniversary, we honored Thomas Alva Edison. With such inventions as the electric (incandescent) light and phonograph as well as a myriad of other inventions, he gave us the joy of *light* and *sound*."

And he continued, "On our fiftieth anniversary, we honored Guglielmo Marconi for his invention of radio or wireless telegraphy. He gave us the joy of *communication*.

"And on our seventy-fifth anniversary, we honored Dr. Robert H. Goddard for his invention of the rocket engine. He gave us the joy of *freedom* in space."

"And now," Dr. Ryder continued in his excited mood, "for our centennial anniversary, we are examining our greatest technological achievement during the past quarter of a century and shall honor its contributor, Dr. Jack Rob

Tains. To do so, we call upon three leaders in science and technology, who have had the privilege of working with our honoree in different phases of his career: his former teacher and advisor, Dr. Joseph Weyl, of the University of Florida; his former colleague at the Communications & Systems, Inc., and currently the scientific advisor to the President, Dr. Ed Davis; and his respected professional colleague in the aerospace field and NASA (National Aeronautics and Space Administration) administrator Dr. Webb's personal representative, Dr. Ned Carroway.''

Having been employed by the Communications & Systems, Inc. (CSI), for the past sixteen years and now serving as president of its subsidiary, the CSI Aerospace Corporation, which he founded, Rob almost always sat at head tables. But he had never been as restless as today. Scanning the array of elegant and enormous chandeliers glittering and coruscating, he wondered about a human injustice: "Who in the world deserves that much recognition as the IEAS is all set to lavish on him?"

Exhilarated by the thought that he had perhaps accomplished something and comforted by the fact that the person who deserved a lion's share of the credit was in the audience, he turned his head to the tables in the honor guests' section and near the head table.

There sat at the first table the dignified, smiling, proud Mrs. Walter S. Tains. Her gray hair was very becoming on her, adding dignity and grace to her almost regal posture. She looked every inch a queen mother. Why not? Her son Rob would be crowned as the centennial contributor to the world of technology.

As Rob's head turned to her table, Mrs. Walter, or Dee as she was known to her friends, was expecting it. She met his glance with a smile as a full acknowledgement for his attention.

He felt that he was reliving his bedtime during his young boyhood when his mother read the stories of great scientists and engineers: Thomas Alva Edison. Guglielmo Marconi. Robert H. Goddard. *I remember them. It was just like last night. Mother made these revelations to me.*

Also, there was Helga McGee at the table next to Mother's. She seemed to attract so much of Rob's attention. Was it because of her white dress? Her interview tomorrow? Her glances at him? She was simply beautiful. Rob looked at her again and again just to be sure that she was real.

The program had a glossy photograph and detailed (two pages) biography of Rob. Helga scrutinized them carefully. She always associated great scientists with the appearance of Albert Einstein: a mop head of silvery hair, a bushy mustache, a face with wrinkles as traces of profound thinking, and a pair of pensive eyes, philosophizing or daydreaming.

Rob was too young and too handsome to fit this image. But at the age of thirty-seven, he was the youngest to be lionized with an IEAS Medal of Honor.

Reading his biography, Helga was amazed, puzzled and awed. But she did not want to dwell on these questions as she was enjoying the pomposity of the occasion and pontification by the speakers. *Heck, I have the whole day tomorrow to ask him all these questions.*

Now, Dr. Ryder had just introduced Dr. Weyl as Rob's former professor, and Dr. Weyl took over the center rostrum.

Full of reminiscences and a touch of emotion, Dr. Weyl painted the picture of a genius in school about Rob's activities.

"He was a straight-A student with an extraordinary sense of ingenuity. He created Mac, the mechanical man or robot, for the student-sponsored engineers fair. Backed by banks of relays, Mac could blink his eyes (with, of course, electric bulbs as eyeballs), turn his head, and raise his right hand to shake hands and greet people coming to the fair.

"It was an instant success," he continued, "and Mac lived on for many annual engineers fairs. No wonder Jack Tains acquired his nickname Rob in recognition of the robot he created."

A man of a small stature and a brilliant mind, Dr. Weyl was Rob's mentor, who had guided him through for his B.S. and M.S. degrees in electrical engineering at the University of Florida, and recommended him to the Stanford University for his Ph.D. degree and then to the CSI Research Laboratories in New Jersey for his career debut.

"Jack 'Rob' Tains. What a handsome name!"

Dr. Weyl then continued in an enthusiastic tempo, "Let me share a secret with you. Except for his mother whose maiden name was Dolores Ryschkewitsch and me who had read Rob's transcript, almost nobody knew Jack R. Tains' middled name was Ryschkewitsch."

Dr. Ed Davis, a former colleague of Rob's at the CSI Research Laboratories and currently scientific advisor to the

President, was the next speaker, and picked up almost where Dr. Weyl left off.

"I first met him when he came to work for the CSI Research Laboratories. We worked in the same laboratory, and almost immediately I was impressed by his intelligence, enthusiasm and creativity in technical work. During staff meetings on technical projects, he was always the one who offered a good proposal. And chances are, his was almost always the final choice, by our laboratory director, as the route we followed."

Adjusting his chrome-rimmed glasses, he continued, "Within a year of his arrival at the laboratories, he made a great invention, and for the decade that followed he was able to refine and generalize it into a theory.

"The invention-and-theory is now commonly referred to as the Human Intelligence Model or H.I.M. The concept of H.I.M. is rather simple. Dr. Tains observed at that time that whenever a human being intends to do something (for instance, parking a car), he always *compares* at each moment his objective (e.g., the final desired parked position of the car) with his observation of the current status (e.g., the current position of the car). The *actual input* to his mind is therefore the difference of the *reference input* (or *his objective*) and *observed output* (or his *observation*) at each moment, and this process is a continuous one. This is what Dr. Tains called a Human Intelligence Model, and he predicted this simple model is responsible for many outstanding human qualities."

Although Dr. Davis had a small build, he continued to feed a sonorous voice into the microphone: "Dr. Tains applied his H.I.M. theory first to electronic circuits and then during the ensuing years to systems with a simple model by comparing the *reference input* and *observed output* (through a feedback loop) with a comparator, and then feeding their difference into the *processer* or *controller*. And the results were remarkable.

"The H.I.M. circuits or systems created by Dr. Tains have been dubbed as *robots*. Unlike his earlier creation, Mac, the robot, which had the appearance of a human being, his H.I.M. robots do not have human appearance and they have various shapes and sizes. But they have many intrinsic human qualities of intelligence.

"In his earlier years of experimentation with electronic circuits, he was able to research and prove that his H.I.M. electronic circuits or amplifiers have varied human qualities which can be described with simple non-mathematical terms. These H.I.M. robots are found to be *broad-minded* (in terms of broad frequency bandwidth), *even-tempered* (in terms of small distortion at different frequencies), *agile* (in terms of quick response to command), *stable* (in terms of no sustained oscillations), *visionary* (in terms of improved signal/noise ratios), and *resolute* and *reliable* (with sustained 'vitality' or 'gain' at a stable level, even if one or more vacuum tubes, transistors or other circuit elements fail to function properly). What is particularly amazing about this is that Dr. Tains has meticulously predicted these qualities mathematically and then verified them experimentally.

"In the subsequent years, Dr. Tains extended his H.I.M. concept, beyond the realm of *electronic circuits*, to *systems*, with other more complex qualities, which he believes are also human but which cannot be described with simple terms except with advanced mathematical formulation."

Proud, pleased and excited, Dr. Davis continued, "Dr. Tains has finally created H.I.M. robots in terms of electronic circuits and systems with amazing human *intelligence* and *sophistication*. And their applications are vast and extensive. Almost all electronic amplifiers now in use are H.I.M. robots, and electronic amplifiers are the workhorses in all electronic systems. And almost all control systems as well as many other systems are also H.I.M. robots under the various names of feedback systems, control systems, fire control systems, servomechanisms, etc."

"For all the applications of electronic amplifiers and control systems, Dr. Tains has literally laid the foundations for the electronics, communications, computer and aerospace industries as well as provided the basic control mechanisms for all other key industries including chemical, nuclear, petroleum, etc."

Dramatizing his presentation, he asked, "What would happen to the industrial world if all H.I.M. robots were taken out of their present stations?" He answered himself in a deliberate low tone, "The world would *stop*."

Dr. Ned Carroway of NASA was the third keynote speaker. In less dramatic but equally effective expressions, he praised Jack Rob Tains for his direct contributions to the aerospace industry.

While a key member of the CSI Research Laboratories in New Jersey working primarily on electronics and communications problems, Rob was able to extend these activities to aerospace applications. Recognized by the Communications & Systems, Inc., management headquartered in New York, Rob received the mandate to initiate, organize and head the new CSI Aerospace Corporation in Long Island, New York. The CSI Aerospace Corporation was now a leading aerospace company in the nation with its major activities in the Apollo program. In addition to his contributions in H.I.M. robots, which were basic circuits and systems in aerospace activities, Rob also contributed technically as well as in the realm of management in the field of aerospace. As a matter of fact, he was a leading spokesman for the aerospace industry.

Dr. Carroway was correct in his appraisal of Rob's contributions, and finally concluded his speech with this accolade, "Thanks to Dr. Tains' contributions in the H.I.M. robots and to his other direct contributions, we may now expect a moon landing by man within a decade. A fantasy? Yes. But we owe the potential realization of this fantasy to Jack Rob Tains."

An enthusiastic audience applauded and stood in respect. Rob stood up and bowed as a gesture of his appreciation.

IEAS President Frank Baker made this Centennial Medal of Honor award presentation—a substantial check, a gold medallion, and a scroll. The presentation was simple and dignified.

Rob walked to the center rostrum and began his speech by recognizing and giving due credit to several people in the audience who had inspired him and had contributed to his education, training and eventual success as witnessed today: his mother, Dee Tains; his former classmate at the University of Florida, Dr. J. V. Atanasoff who was later to be recognized as the inventor of the electronic computer (*Datamation*, February, 1974, pp. 84-89); his teacher, mentor and trusted friend, Dr. Joe Weyl; and his Ph.D. adviser at the Stanford University, Dr. Frank Terman.

Rob's speech was brief. After a review of the recent progresses in technology, he expressed his pleasure that his H.I.M. robots were making substantial contributions to the well-being of mankind.

He thanked the keynote speakers and in particular Dr. Ed Davis for his human characterization of the H.I.M. robots as *even-tempered, broad-minded, agile, stable, visionary, resolute* and *reliable*.

And he concluded his speech: "I am extremely pleased that the robot has learned so many of the human being's most noble and sophisticated qualities. But, let us re-examine the behavior of our own human society. Do we have a lawless, prejudicial, cumbersome and unstable society with a lack of vision, courage and reliability? Is it time that we relearn some of our original fine attributes from our 'robot' friend and re-establish a society based upon logic and justice before our civilization crumbles on us?"

Rob received a standing ovation. He wasn't sure whether it was because of his honor or because of his words.

Two

Helga arrived at Dr. Jack Rob Tains' office in the Manhattan Headquarters Building of the giant Communications & Systems, Inc., promptly at ten in the morning.

"Ms. McGee?" the receptionist in the outer office asked as Helga approached her desk.

"Yes."

"Dr. Tains is expecting you," she said. "Will you come in with me, please?"

As they walked into Rob's spacious corner office, Helga scanned her eyes on the setting surrounding the man she was to interview. On the thirty-fourth floor, the windows had a commanding view of New York's skyline with a distant, but majestic scene of the Hudson River in the background. The massive modernistic mahogany desk had only a calendar, a marble pen stand, a memo pad, and a model of a sort. *A clean-desk man and good organizer, eh?*

As Helga approached his desk, Rob stood up and extended his right hand.

As they shook hands, she said, "Congratulations, Dr. Tains, for the tremendous honor last night!"

"It's very kind of you, Ms. McGee."

He pulled out from his desk and led her to the sofa. They sat down as the receptionist left the room.

Helga looked and smiled at her host. "Finally, I've made it. I am interviewing the Centennial winner of the Medal of Honor award!"

Rob was somewhat at a loss for words. "Ms. McGee—"

She interrupted rapidly, "Call me Helga. Everybody calls me Helga."

"Yes, Helga. Then I am Jack or ——"

"You are Rob. Your biography said so."

He burst into a good laugh. "Do I really look like a robot?"

"I must conclude that the robot is a good-looking species." And she chuckled.

"Thank you." He was pleased with the answer.

For more than an hour he fielded a barrage of questions, reviewing the details of his biography which she had read in the awards program last night and expanding them. And she also dutifully took notes for the piece she was assigned to write for the *New York Tribune*.

She learned a lot. Rob's father was professor of physics at the Columbia University for forty-two years, retired some ten years ago, and died five years later. Rob was the only child and always loved science and mathematics, and he also loved to tinker with gadgets. They lived in Washington Heights where his mother still lives.

Helga also dug somewhat into Rob's present status. As president of the CSI Aerospace Corporation in Long Island, he normally spent the major portion of each week in Long Island in the executive office or in the laboratory (as laboratory work is still one of his loves). For a day or two each week and occasionally for all five days of the week, he spent in the Manhattan headquarters of the parent organization, the Communications & Systems, Inc., time for meetings, planning sessions, meeting contractors and clients. He also traveled a lot throughout the country and around the world, and also on brief visits to the CSI Research Laboratories in New Jersey for consultation with his former colleagues of many years.

She also learned that Rob was a bachelor at thirty-seven. Such an eligible bachelor! Why did he remain unmarried? An interesting topic for investigation and reportage to the *New York Tribune*! Despite her inquisitive mind and her intensive interest in this matter, Helga refrained from probing into the subject in an exhaustive fashion, refrained at least for the time being.

And she continued her interview.

"Rob, besides you intensive love for work, what are your other loves?"

"I think, water. I always love water."

"Why?" She was curious.

Rob looked quizzical. "Helga, I cannot tell you exactly

why. Maybe, it is the setting with water that fascinates me.
Maybe, it's the form and expression of water."

As if mesmerized by the very thought of water, he
continued, "When I was small, I used to walk to a small park
in Washington Heights in the shadow of the Washington
Bridge overlooking the Hudson River, and sat there for hours.
As the water was flowing in the mighty river, so was my
thought process. It was the beginning of the formation of my
philosophy of life and my ambition. As nature provides us
with resources and power as symbolized by the water, I feel I
should dedicate myself to science and technology, utilizing
and harnessing nature's gifts for the welfare of mankind."

"Rob, have you been to the Washington Heights park
lately?"

"No, I have not. Ever since I was five, I have also been
going to the family hideaway on the Long Island Sound. My
father bought the beach property, some 100 acres, and we
went there during the weekends and lived there for the entire
summer, year after year. Except for my college and graduate
school years at the University of Florida and Stanford
University, I have been going there or living there for a span
of thirty-two years.

"At the hideaway," he continued his monologue, "I used
to sit over the bluff or a boulder overlooking the Long Island
Sound, watching the tranquility of the water and the sky
with changing cloud formation, and musing over the vicis-
situdes and immutable truths of this human world.

"The choice of the University of Florida and Stanford
University for my education," he continued, "was in fact
partly because of my love for water and partly because of the
quality of their educational programs. When I was at the
University of Florida in Gainesville, I often drove to Cape
Kennedy, then known as Cape Canaveral, watching the vast
expanse and serenity of the white sand, boundless ocean, and
blue sky with daubs of white clouds. And when I was at the
Stanford University, I drove along the Pacific Coast up and
down, and in particular I loved the Monterey Peninsula where
I watched vigorous waves pounding on the rough seacoast.
The rocky coast with trees which look almost artificial in
their shape to an Easterner, invigorated by roaring waves,
symbolizes courage and fortitude."

Helga was fascinated and listened attentively. And she
asked, "Rob, was there any concrete impact of these bodies
of water upon your life?"

Reflecting for a short moment, Rob grinned. "Yes, it was

due to my reflection on the hideaway beach during a weekend when I was working for the CSI Research Laboratories that I conceived the Human Intelligence Model, a concept which is reputed to have revolutionary impact on technology and our daily life."

"Rob, what was in your mind when you discovered the H.I.M.? In other words, how did you discover it?"

Rob replied, "When I watched the gentle waves slapping at the five rocks (incidentally, these are my favorite rocks) which were half-submerged on the hideaway beach, I heard the rhythm and realized that it was the simple repetitive motions that made up the laws of nature prescribing the behavior of the Universe. Then it was only logical to surmise that it must be some simple repetitive human mental process that prescribes human behavior and wisdom. Then it struck me on how I parked my car in a space on a busy street of a nearby township that afternoon. It was a continuous 'mode of comparison,' comparing at every moment the final *desired parked position* (which I later technically described as *reference input*) with the *current observed position* (technically, as *observed output*) of the car and feeding this difference of comparison into the mental process. This is the concept of my Human Intelligence Model."

Elevated with his own dissertation, he continued, "Helga, I don't think that the human being is a supreme being with an innate complex mental mechanism. Rather, it is the simple concept of the H.I.M., applied repetitively and continuously that is responsible for many human qualities and wisdom.

"Within a few weeks of the conception of this H.I.M. concept, I was able to use a comparison mechanism or feedback mechanism in a vacuum-table amplifier to make it stable and having many other desirable qualities. Later, this concept was applied to transistor amplifiers and integrated-circuit amplifiers. As these stable amplifiers or H.I.M. robots are the workhorses of electronics with extensive applications in electronics, biomedical, communications, computer, and aerospace industries, they are the very foundations of these industries."

"Rob, do you suppose that you'll ever teach me all the principles and scientific know-how about these H.I.M. robots?"

Rob winked, "I'll do it in the days or years ahead."

Helga also winked. Only they themselves understood what they winked about.

While asking all sorts of questions about Rob's career, contributions and life, Helga just couldn't help wondering about the white model on Rob's desk.

"What's that on your desk?" she asked, pointing her finger at the white object.

"That's the model of the Apollo rocket and spacecraft which in a few years will take the first human being to land on the surface of the moon."

Helga looked incredulous. "To land on the surface of the moon?"

"Yes," Rob answered briskly.

"How much do you have to do with it?"

"Plenty. My company is involved to its neck."

Raising his right hand above his head, he said, "I myself am that much involved."

"Rob, when you get to the moon, will you stake a plot on the moon for me?

"You bet I will," he answered with a broad grin. "Helga, let's come down to the earth for a little while, do you want to see my company and the hideaway on Long Island?"

"Yes, very much so."

"It takes two hours of driving to the company. Let's have lunch now."

They ate in the executive dining room of the CSI Headquarters Building. Actually, it is one of the many exquisitely decorated small dining rooms. Very cozy.

The afternoon ride to the CSI Aerospace Corporation was a long one in Rob's gold-colored Porsche, a low, swift and comfortable ride with interesting conversations.

At one point, Rob turned to Helga, "You have been asking me questions for half a day. Do you believe in reciprocity?"

"Yes. And it's only fair. I'll reciprocate. Now shoot!"

Rob was also a good cross-examiner.

Helga was the younger of two children, her brother Steve being fifteen years older. Her mother died during childbirth, and she had never seen her mother. She was raised by the housekeeper Anita, or rather Aunt Anita as she used to call her.

Her father Frank Delano McGee was an architect and a very successful one, with a firm in New York City and heading gigantic urban projects in New York, San Francisco and Atlanta. He commuted from his home in Tenafly to New York daily, and also traveled a lot. Although Helga was

Anita's charge during the weekdays and transported around
by her to school, dancing classes, music lessons and wherever
she wanted to go, Frank made a point to be free of
professional activities during the weekends. He was always
available to be with Helga, seeing movies and plays, visiting
scenic and historic places, and going after the best foods in
New York and elsewhere. They talked a lot and were very
close to each other.

"Helga, what kind of a person is your father?"

"He is feisty, independent and charming," she said with
pride and gratification.

"What have you learned from your father?" he asked.

"Plenty. I learned how to be independent, question the
society and follow my heart."

Rob was curious. "What does it lead you to?"

"Four years of journalism study in Missouri after high
school in Tenafly. Three years in the Bay Area as a reporter
for the *San Francisco Examiner*. Two years in Washington,
D.C., for the *Washington Post*. Now, I am working for the
New York Tribune and taking a stab at the assignment of
finding out about the robot or the man? Who is the master of
the world?"

Rob burst into a hearty laugh. "Good luck, my friend!"

The CSI Aerospace Corporation had an enormous premise
with a campus-like setting on Long Island. Rob pulled in the
car to his parking space and escorted Helga to visit his office
briefly and then to see laboratories, test and development
areas, and manufacturing quarters.

They first stopped at the missile areas, which was
unrelated to the Apollo program. They are sort of "odd jobs"
for the company under contracts with the armed services.

Then they came to the communications, guidance and
control, and electronics areas. They are the heart of the
support areas for the Apollo program. Rob took pains to
explain their activities in layman's terms. Helga was immensely impressed and nodded her head repeatedly.

In one of the electronics laboratories, the microelectronics
laboratory, Rob showed under a microscope an integrated
circuit (IC) to Helga and said, "Helga, here I want to show
you the rapid advances of technology."

Turning to a self-contained slide-projector-and-viewer box,
he turned on the first color slide, narrating, "This is a *vacuum
tube* as the basic workhorse for electronics during the past
years. As shown in this slide, it compares in size to an
orange."

He then moved on to the next slide and continued, "Here is a *transistor* which was invented as a revolutionary development some years later to replace the vacuum tube as the workhorse for electronics. In this picture, we see how it compares with a peanut (without shell) in size."

He then turned to the third slide showing a "quarter" coin put side by side with a silicon wafer of similar size and shape. And there were grids on the wafer dividing it into 300 square chips.

Rob then continued, "During the past several years, we have another revolutionary development. The integrated circuit, as you just saw with the microscope, was invented. To appreciate this new invention, just imagine that the small chip, 1/300 of a quarter coin in size as shown on this slide, is not just equivalent to an orange. It is actually equivalent to, say, fifty oranges in addition to fifty apples and fifty bananas."

And Rob continued to explain, "A vacuum tube or a transistor, as figuratively compared to an orange or a peanut in size, is merely a simple device. An integrated circuit is a circuit or a system, consisting of many, many simple devices, say, the equivalent of fifty vacuum tubes (or oranges) and fifty resistors (or apples) and fifty capacitors (or bananas) complete with interconnections to make it function like a circuit or a system."

Rob then snapped the projector button and showed several slides of typical integrated circuits enlarged on the screen.

Fascinated by the designs of these integrated circuits, Helga asked, "Are you sure that these are not prints of modern art?"

He answered, "So you are admitting that the engineer also has some artistic sense."

He then continued on with the additional slides, comparing completed integrated circuits with a fingernail, and with a needle.

"These integrated circuits are incredibly small as we have shown, use very small electric power and are unusually reliable." And he continued, "Think about the implications of its potential applications in electronics, communications, computers, and medicine, and in particular, in space exploration where size and weight of our communication, instrumentation and control packages are the overriding considerations."

As she listened, she also gasped. Was this reality or science fiction? Looking at Rob's handsome face and his broad grin, she had reasons to believe that all was real.

The entire visit to the CSI Aerospace Corporation was a high note for Helga.

After they left the highway, they drove through tree-lined or tree-shrouded winding roads. And on the property of his own Hideaway, the road meandered through a grove of mixed tall and small trees—maples, hazels, tulips, oaks, cherries, and dogwoods. When they came out of the shade, they passed through a slit of the bluff to the beach. The bluff, as a long hilly embankment along the beach, was covered with patches of green vegetation and some small trees, and bare dirt offered as walk paths leading to the top of the bluff, where you have a command view of the Long Island Sound. It is the widest portion of the Sound, and all you can see is a body of water, no land on the opposite side.

Along the bluff on the beach was a long white clapboard framehouse. It looked as if recently painted—neat, lovely, but, decidedly not a new structure.

Coming out of his Porsche, they came on the beach. For a moment, they stood still, sharing with the surrounding its serenity.

"Helga, this is my hideaway."

"It is a beautiful place. A perfect setting to fall in love with nature."

"I love it," he said in a low voice so as not to disturb the tranquility. "I spent the summers of my childhood and many more days since then, in this white house, on this beach, atop the bluff and in the maple hazel grove. I believe that they have shaped my views, philosophy, convictions and and beliefs."

Helga nodded in silence.

Pointing at the five rocks half-submerged in water, he said, "These were my *five continents* in my make-believe world during my childhood days. I used to jump from one to another, pretending I was a world traveler. And these were the same rocks slapped by gentle waves that taught me the concept of the Human Intelligence Model."

As he was talking about his childhood and earlier experiences, his polluted nostalgia swam into focus. He saw himself barbecuing and picnicking with his parents in the maple-hazel grove, bathing and frolicking with some occasional friends in the water, collecting rocks and shells along the beach and bluff (way beyond their own property), and watching boats passing by.

"Rob?"

"Uh—yes, Helga." He collected himself and said, "Let's go inside the house."

Except for a fresh coat of paint put on only a few weeks ago, the appearance of the house had not been changed since Rob's father bought it some thirty-two years ago. However, Rob had remodeled its interior completely during the past few years.

"Rob, this is perhaps the newest old house on the entire Long Island! And I like the way you refurnished the house."

Cushy, comfortable, and Scandinavian modern.

A large piano sat in the corner. Helga walked over, sat down, and played some of the old songs she learned during her high school days when she was taking piano lessons.

After a lobster dinner at Rob's favorite seafood place on Long Island, they drove back to Manhattan.

As they were approaching her address, Rob asked, "Helga, how about having dinner with me on Saturday, that's the day after tomorrow?"

Sizing up him with a glance, she chuckled and asked, "Why?"

"Because," he said.

"That's a good reason. Besides, I need more interviews to make a living."

"I'll pick you up at six-thirty."

She smiled at him. "Alright, six-thirty."

When he took her back straight to the door of her apartment, she planted a gentle good-night kiss on his lips, and said, "Rob, it has been a terrific day. Thank you very much."

And then she shook his hand, giving him a terrific grip, and winked and bade, "Good night, Rob." Half an hour after the handshake, Rob could still feel the grip.

Three

Sixteen days had passed since the interview. During these sixteen days, they had ten lunches, eight dinners, and six Broadway plays and musicals.

They also drove to many scenic spots, where Rob used to visit, watching the water and its varied forms. They went to Fire Island, Montauk Point and Orient Point on Long Island and the Connecticut coast of the Long Island Sound, the lighthouses, the sunsets, the picturesque island setting, the sea grass on sand dunes, sunny skies and refreshing breezes, sparkling water and white sand, the blue Atlantic at its best—for Rob, they were vignettes and memories of his boyhood. For Helga, it was a new dimension of learning and sharing Rob's experiences. And she was mighty happy about it.

It was a beautiful Saturday. And it was a very special date for Rob and Helga, for he had invited her to a gourmet dinner at his Hideaway. And he was the chef.

Helga had learned that Rob was a connoisseur of Chinese food. Only last weekend, they went to the Chun Cha Fu Chinese Restaurant on about the ninetieth block of Broadway, where they served tea-luncheon (a Chinese custom) with special Chinese hors d'oeuvre and pastries of different cuisines (representing different geographical regions of China) on Saturday and Sunday noons. Only a gourmet of Chinese food would know how to order. But Rob seemed to be an old hand. They skipped those standard dishes like wonton soup and egg rolls for the "foreigners," and ate what the discerning Chinese would eat: lo mien, fried dough (in elongated, twisted form), bean milk (which Helga didn't like), steamed buns of varied kind, fried dumplings, and braised stuffed Chinese mushrooms (which were not on the menu but were a special just for that day). They really ordered too much to be consumed. But, they were meant to be samples for Helga. Hurrah for Helga! In addition to her main piece of the interview, she wrote another article "Mr. Robot and his Chinese Hors d'Oeuvre" for the *New York Tribune*.

For the Chinese dinner this evening, Rob had prepared for days. He bought all the necessary ingredients from Chinatown, and did in advance most of the preparatory work on the Peking duck course as well as much of the cutting and slicing as required for Chinese cooking. Helga had promised to help, and Rob took her offer.

Helga drove to the Hideaway about five, and joined Rob for the preparation of the feast.

Although she did not know what she was doing, she dutifully followed Rob's instructions, cutting, slicing, and almost constantly washing pots and pans.

"Is this customary to wash pots and pans while you cook?"

"Yes," he said. "You have often to cook different ingredients separately and then cook them together. That's why Chinese cuisines are so different in flavor and taste."

"Have you cooked Chinese dinners many times before?" She was curious.

"Yes, I am a weekend Chinese gourmet cook."

"Since when?"

"Since my University of Florida days," he answered. "When I was studying there, I took an evening course known as 'Chinese Culture As Expressed in Cooking,' taught by a Mrs. Dorothy Chen and sponsored by the local community college."

"Did you really pass the course?" She was reasonably sure that he did. But, she thought it was interesting to pursue.

"You bet I did." He was very proud, pointing to his certificate on the wall. For a person of his distinction and numerous honors, that was the only plaque on his wall. Strange, isn't it?

For almost two hours since her arrival, they worked. Finally, the dinner was ready.

Following meticulously the way formal Chinese dinners are served, they started with two cold plates, sliced braised star anise beef and smoked fish. (Rob didn't cook them; he bought them ready-cooked in Chinatown.) And they drank and toasted to each other with Lichine's Rosé D'Anjou wine.

Then they followed with the hot and sour soup, a Szechwan cuisine popular in Southwestern China. Boy, was it hot with all the spices! But both of them liked it.

Peking duck followed along with the pancakes, sauce and green onions. Crisp duck skins, juicy meat, the sweet but somewhat pungent sauce, the green onion bulbs cut into "flower" shape—all were put into the pancake as a package. What a succulent, delicious and exotic package! It defies description.

"Rob, what do you think about when you eat Peking duck?"

Rob thought for a moment and answered, "It was exotic! And I'd like to visit the lands of the exotic. Peking is one of them. Think about all the Peking duck we can eat."

They laughed. And Helga proposed a toast, "Let's drink to Peking for its great culture and its great cuisine!" And they drank.

Knowing that Rob had been to the various European countries, Helga asked, "What are the other exotic lands you'd also like to visit?"

"Russia and Venezuela."

"Why?"

"I have seen the picture of the St. Basil's Cathedral of the Kremlin many times. It's so exotic in every way—its color combinations, its soft ice-cream-cone-shaped domes, its architectural style, and its majestic setting on the edge of the Red Square. And I fell in love with it at the first sight when I saw its picture."

"You are the romantic type—aren't you?" she asked.

"Well, the St. Basil's Cathedral is only symbolic of what Russia stands for," he said.

"What do you know about Russia?"

"I know very little about Russia. I might as well say that we, collectively as a people, know very little about Russian people. But, I suspect that behind the enigmatic facade of the St. Basil's Cathedral there lies the richness of a magnificent culture, and that behind an image of mystery and exoticism (at least to us) there lives a people who are colorful, intelligent, kind, peace-loving and just as courageous as we are."

Helga applauded. "Bravo! It's a great speech."

Rob took a bow and grimaced. "I'd surely like to visit Russia someday."

Rob then went back to the kitchen and Helga followed.

And they managed the last four main dishes with Mrs. Dorothy Chen's recipes (which Rob kept and used for years): shredded meat with eggs, sweet and sour pork, stir-fried chicken with nuts, and shrimp with lobster sauce.

As they were eating, Helga asked, "Why do you want to visit Venezuela?"

"I have only been exposed to the English-speaking culture. And I'd love to have some exposure to the Spanish-speaking culture. Besides, I have had a standing invitation to visit Venezuela for twenty years."

"From whom?" she asked.

"From my former roommate at the University of Florida and my best friend, Dr. Jesús Vivaldi Casanova, who is now the dean of engineering at the University of Carabobo in Valencia, Venezuela."

Helga smiled. "Rob, I wish you luck for your exotic dreams."

They concluded their dinner with almond float as desert. And they enjoyed tea and classic music in relaxed and languid luxury on the sofa—Beethoven, Bach, Mozart, Tchaikovsky.

They came out of the house and walked on the beach hand in hand long after the sunset. The moon was full and illuminated the water in distinct contrast to the land (namely, the beach). It also silhouetted the bluff and trees.

Rob and Helga took their shoes off, and walked in the water. Although it was in June, the water was still chilly but very refreshing. Finally, they ended up on the five rocks. There they stood as if in a still picture, not moving, not a sound uttered. They just looked at each other, smiling with dreamy eyes.

Rob was also counting his blessings on these five rocks. Nineteen years ago, they inspired him for his epic discovery. Now, he was wallowing in a sensation unknown to him before. She was beautiful. She was lovely and gentle. And she was intelligent. To make sure that she was real, he held her arms with his two hands and held her close enough to feel her warm breath.

After a long spellbound time interval, Rob held Helga closer and kissed her. And they embraced and kissed more. First, they moved their faces against each other and then kissed over a long breath.

Finally, Rob broke the silence. "Helga, I am in love with you."

"Rob," she whispered. "I am in love with you too."

After a succession and repetition of looking at each other, embracing and kissing, Helga held Rob's two hands together, enveloped them with her hands and gave a terrific squeeze. And she said, "Love is not a comma, but a period. A helluva grip of a period."

And what a grip! Rob grimaced. Helga looked up and saw a curious look on Rob's face. And she was anxious to explain, "It was very simple. When I was a fifth-grader and asked by the teacher to tell the difference between a comma and a period in class, I constructed a sentence as an analogy: 'Love is not a comma, but a period.' And it brought down the house with laughter. But, I was proud of it."

Rob gave a solicitous smile and Helga continued, "I meant that love is not a pause like a comma, but a completion and ultimate state as implied by a period. And I thought it was cute."

Apparently, her interest on this matter was immense and she continued her monologue. "That evening I tried it on my father and also squeezed his hand as I always did. My famous expression now came into existence. It was a bit childish, and it was a bit foolish. Perhaps both. But, it gave me years of pleasure during my childhood in expressing my love to my father. You know, I just adored him."

While he nodded his head, he seemed also to be inquisitive. And Helga smiled, "Tonight is my first adult experience in using this expression."

Now, she again squeezed Rob's hands with a strong grip. With a slight contortion on his face, Rob smiled and chimed in unison with Helga, "Love is not a comma, but a period. A helluva grip of a period!"

After a long period of gentle and loving expressions, Helga suddenly broke off, left the five rocks, and jumped back to the beach.

There she undressed, ran and dashed into the Long Island Sound.

Rob followed. He also undressed and rushed into the water.

They chased each other. They swam. They frolicked. And they laughed.

When their energy was spent, they quietly stood in the water waist-deep, holding each other dearly.

Finally, they came out of the water, wrapped themselves in a blanket, and sat on the beach.

They did not talk. They were too happy and contented to

talk. They did not want any change. They just sat there to wait for sunrise.

Four

It was not surprising that Rob and Helga were married by the sea; that is, by the Long Island Sound at the Hideaway.

It was a simple wedding. Only members of the two families and Rob's closest associate, Executive Vice-President Jack Rosier, and his wife Mary, were invited.

Rob's mother Dee was extremely pleased. For years, she tried to talk Rob into getting married. She almost had given up the hope of ever becoming a grandmother. And now the prospect had brightened.

The bride's side was well represented at the wedding: Father Frank, Aunt Anita, Brother Steve, his wife Ann, and their lovely twin children, Charmaine and Michael, both eleven and Helga's favorites.

The wedding was almost like all the other weddings except that it was held at 7:00 A.M. and that Helga, Rob and their minister and friend Sam Holyhouse wore no shoes. Helga wore a long white dress, Rob his tuxedo and Mr. Holyhouse his robe; all were properly attired.

Helga, Rob and Mr. Holyhouse stood slightly in the shallow water, more or less on the edge of the water. All the wedding guests stood more inland on the beach watching the ceremony.

Through their wet feet in the sand and in the water, Helga and Rob felt that they were in touch with nature. And the morning fog made the Long Island Sound even more mystic than usual.

"Friends," said Mr. Holyhouse, "we have been invited here to witness the union of two people in marriage. They have chosen these surroundings in nature. For its serenity. For its beauty. For its strength. For its changes in day and

night. For its changes in high and low tides. For its changes in
rain and shine. For its changes in wind and storm. And for its
perpetuity in these changes that symbolize this union."

As Mr. Holyhouse continued his sermon, Helga and Rob
watched each other and sought meanings of every word
uttered by their minister.

A pair of sea gulls flew by and caught their eyes. They
smiled at each other. It must have been a good omen.

And Mr. Holyhouse continued to preach the sanctity of
holy matrimony. And then he came to the questions of
commitments, of which both Rob and Helga said in turn, "I
do."

Then they exchanged rings and recited the marriage vows,
taking each other till death do part.

Suddenly, a strong breeze came from the sea. In its wake,
the sea grass and shrubs seemed to nod approvingly. And
sun broke through the fog smiling blissfully.

It was no surprise that they went to Venezuela for their
honeymoon. Immediately after the ceremony and reception,
they headed for the John F. Kennedy International Airport.

It took their Pan American Boeing 727 four and a half
hours to arrive at the Maiquetia Airport. Rob's best friend
Jesús Vivaldi, who lived 120 miles away in Valencia, was very
considerate. To respect the privacy of the newlyweds, he did
not come to greet them. Instead, he sent his chauffeur José
along with his Mercedes to greet the Tainses and take them
to the Hotel Tamanaco in Caracas. Fortunately, José spoke
reasonably good English and was also to serve as a guide for
Rob and Helga. José and the Mercedes would be at Rob's and
Helga's disposal for their entire stay in Venezuela—their
sightseeing in and around Caracas, travel to Valencia (to visit
Jesús) and to the Andes country, and then return to Caracas
and Maiquetia Airport for the return flight to the United
States.

As José drove them from the Maiquetia Airport to Caracas,
Rob and Helga were impressed by the winding modern
highways in the mountains. Reaching into an altitude of
3,000 feet at Caracas from the sea-level Maiquetia on an
ascending journey was a pleasant experience. And the high
altitude is obviously responsible for the excellent climate that
Caracas enjoys year round. As they drove along and entering
Caracas, Rob and Helga marveled about the panoramic view
of the famous city—majestic mountains, beautiful gardens,
modern structures and skyscrapers, and crisscrossing express-

ways as the arteries holding the human civilization (or in a less grandiose term, the human masses) together like an octopus.

The Hotel Tamanaco was a pleasant surprise for the Tainses, a magnificent hotel and tropical gardens on the hills of Las Mercedes. When they entered the reserved bride's suite, a gorgeous flower basket (a Venezuelan custom?) was sitting on the dresser—beautiful and glorious fresh flowers with a note signed, "Congratulations and best wishes from the Jesús Vivaldi Casanovas."

Seeing the note, Rob asked, "Isn't Jesús sweet?"

Helga replied, "I just love him."

For the next two days they never left the hotel. They were in the gift shops downstairs only once, where Helga brought a saffron hand-knit poncho for her niece Chairmaine and a gold-coin medallion for her nephew Michael.

They were either at the kidney-shaped giant swimming pool or secluded themselves in their suite.

In the boudoir of their suite, they enjoyed the tenderness and intimacy of their new relationship. They explored the contour of their bodies and learned new meanings and dimensions of their love.

Rob had not known and suspected that she could be so sweet, tender, and loving. But he was grateful for that.

As she rested her head on Rob's chest or her hand was holding Rob's in a constant grip, they had their chitchat: her crush on her science teacher, his first date, her senior prom, and a montage of earlier encounters with boys and girls and Rob's robots, grins, laughters, and a few faint notes of jealousy.

While they were talking about Rob's robots, Helga was curious. "Rob, are there sexes among the robots?"

"Yes, of course. Let's consider the transistor amplifiers as robots. The transistors are either of the NPN or PNP types; they have exactly the opposite polarities and are hence different sexes."

"Rob, what do you do with them?"

"In a complicated circuit or system, I use NPNs and PNPs in succession. In other words, I 'pair' them in order to achieve sex harmony and make the whole thing work."

Raising her head to his chin and still leaning against his chest, she asked with a mixture of disbelief, "Rob, have you ever screwed a female robot?"

Kissing her forehead as a gesture of assurance, he answered, "No, I have not."

"Why?"

"They wouldn't have me. You know that the robots are a bunch of damned racists."

"So you have to take the second best?"

Looking at her with a grimace, he said, "Precisely." And they burst into big laughter.

They seemed to intersperse their preoccupations in the bedroom with relaxation periods at the poolside. Occasionally, they had a few dips.

While at the poolside, they would sit or lie down on some reclining chaise lounges, sipping gin 'n' tonics or Tom Collinses. The staircase architectural-style hotel building on one side and a flagpole-lined perimeter on the other gave them a sense of security and privacy. With a giant pool as its centerpiece, this domain was further decorated by palm trees and tropical plants; and to New Yorkers like Rob and Helga, there was an insouciant tranquility.

Beyond the flagpoles, there was the majestic view of the mountains in the background, somewhat shrouded by fog adding to the mystery of the spectacular landscape.

Rob and Helga ventured out of the hotel on their fourth day in Venezuela, sightseeing in Caracas: the Central University, the Botanical Garden, the Plaza Venezuela. At noon, they ate pizza at a restaurant in the giant Chacaito Shopping Center where Rob bought some gold pins and a modern watchband-like gold bracelet for Helga. (Venezuela is famous for its gold jewelry.) The afternoon was consumed in more sightseeing: Plaza Bolivar, the Capitol with its gleaming gold dome.

Their fifth day in Venezuela was their last day in Caracas. It was a very special day, for they were to visit Mount Avila.

They arrived at the cable-car terminal at Cota Mil and Avenue Principal Mariperez right after lunch, and waited in the covered corridor of this modernistic terminal for half an hour for their ride to the peak of Mount Avila.

Soon it was their turn, and they entered the orange-colored cable car facing backward toward Caracas. As the cable car moved, they witnessed Caracas in a grand aerial view: a great city contained in a basin surrounded by magnificent mountains. As they rose higher and higher, they had the sensation of sitting inside a reflex camera zooming out of the great metropolis behind them. And what they saw yesterday—the university with a mighty red wall (for one of its buildings), the staircase-shaped Hotel Tamanaco, the twin towers of the Plaza Bolivar—became smaller and smaller. Rob took a few pictures with his camera.

Since Venezuela is a Spanish-speaking country and nobody (perhaps) would understand their conversation, Rob and Helga felt a new freedom and took advantage of it.

As their cable car was moving, Helga asked, "Hi, space expert, are we really heading for the moon?"

"Yeah," Rob answered snappily and with the confidence of an astronaut. "And we are getting out of the earth's atmosphere."

"Do you feel weightless?"

"Yes—particularly with you around here."

"What do you want to do on the moon?" she asked.

"To stake out a mountain there and call it Helga's Period."

"Why?"

"Because it is a period, the ultimate limit. And nobody can surpass it."

As she said "Really?" Helga held Rob's right hand with both her hands and gave a terrific squeeze. And he grimaced.

As their cable car ascended further, the ever-diminishing Caracas landscape suddenly disappeared behind the cloud formation. Or was it merely fog?

In a few moments, the car reached its terminal atop the 7,052-foot Mount Avila, an ultra-modern structure of steel, masonry and glass—with a large portion of it in glass. Rob and Helga walked around the terminal and visited the skating rink. And as they walked outside, they saw a giant floral clock with its arms clicking along.

"I wish I had the magic of having my wish granted," Helga said.

"What is your wish?" Rob asked.

"I wish I could stop this clock, this world and our life. I am happy. And I just love this moment."

"Would a perpetual calendar of happiness do?"

She wouldn't take any chances and said, "Only if you guarantee it."

"Sold!" Rob snapped.

Atop Mount Avila, there is a road leading from the cable-car terminal to Hotel Sheraton-Humboldt, a white-tower structure and famous landmark in the Caracas area. Rob and Helga walked along this road and stopped many times. On their right was Caracas with its mountain walls and on the left, the valleys and mountains facing the Caribbean. They also sat on the low walls along the road. At an altitude of 7,000 feet above sea level, the world was literally at their feet. As a matter of fact, it was physically the highest point reached by Helga; she had only been in the Catskills (peak of Mount Marcy at 5,344 feet) and the Smoky Mountains (Clingman's Dome at 6,643 feet). They looked in all directions and peeked through some tree branches, wondering at the grandeur of nature.

Along the way, they saw a flower vendor. Rob bought a bundle of bright-colored fresh flowers and presented them to Helga.

Before reaching Hotel Sheraton-Humboldt, they left the main road, and followed a trail on the left, walking toward the valley. There were trees and underbrushes on both sides shading the trail. The scene was mysterious, quiet and lovely. And they stopped to enjoy the scenery several times.

Suddenly the path opened up and there was a small plateau in an open space. Or rather, a hump. And Rob yelped, "Bravo, I have discovered my mountain on the moon!"

They cleared up the patch of land with their hands and used tree branches as brooms to remove fallen leaves and debris. Now, the moon dust was showing.

With a stiff tree branch, Rob drew a giant heart and wrote on it "Helga's Period." And then, raising his right hand as if taking an oath, he intoned, "As the supreme robot of the world, I declare thee *Helga's Period*!" Helga then gently laid her flowers in the heart neatly below the inscription, and concluded their ritual.

They sat at the Helga's Period, holding their hands in firm grips and gently kissing each other. And they sat there for three hours.

They finally made their way to Hotel Sheraton-Humboldt, where they had dinner, with a commanding view of the mountainous country. After the meal, they leisurely walked back to the cable-car terminal. It was sunset in Mount Avila. The terminal building with its modern profile silhouetted in the diminshing sun and a few scattered clouds in reddish purple dotted the darkening sky.

A cable car now flashed into the scene hanging above the low clouds. In the magnificent view of nature, Rob and Helga felt the insignificant presence of their physical beings.

However, they also felt the ubiquitous presence of their love and they could feel it, smell it and touch it almost everywhere.

On the sixth day in Venezuela, they left Caracas and headed for Valencia in Jesús Vivaldi Cassanova's Mercedes. It took José two hours and ten minutes to drive there.

They checked in at Hotel Intercontinental, where Jesús had made reservations for them. Rob and Helga were surprised—pleasantly surprised—by the elegant hotel. A valley seemed to have been created for this hotel, and a network of enormous swimming pools, cabanas and porticos blended harmoniously with the sloping foothills around the resort.

They had a lovely evening with the Vivaldis—first, dinner at the elegant Club Hipico and a quiet long visit at the Vivaldi residence. For Rob and Jesús, it was to catch up about the happenings during the past nineteen years. For the women-folks, Helga and Jesús' wife Gladys, they were just fascinated by listening to their husbands.

Their three-day stay in Valencia was highlighted by their visit to the *corrida* (or bullfight) at Jesús and Gladys' invitation at Plaza Monumental. It was a new exciting experience for both Rob and Helga. As they watched the picadors on silent horses (with vocal chords cut) weaken the bull by piercing its neck muscles with lances, *banderilleros* confront the animal and thrust barbed sticks between its shoulder blades, and the torero goad and kill the bull, they were excited but were also seized with a sense of guilt.

After they left Valencia, they traveled to the desert city of Barquisimeto, where they had an Argentinian meal, mostly roasts and barbeques at Parilla Argentina, a new experience for the Tainses. Then they traveled along to the Andes region in the Merida State where they admired breathtaking panoramas of snow-capped majestic mountains, green valleys, cascading rivers and still lagoons.

They did not return to Caracas. Instead, they traveled to Maracaibo, center of the nation's oil wealth, where they took a New-York-bound airliner to return to the United States.

Five

Two winters had passed since they were married. And it was spring in New York again.

They actually had three homes: an apartment in Manhattan, an apartment near Rob's office in the CSI Aerospace Corporation on Long Island, and their Hideaway on the Long Island Sound.

They both worked hard. Rob's Apollo project alone was going at a frenzied pace. And Helga's assignment at the *New York Tribune* took her all over the town. But they almost always managed to get together Wednesday evenings in their Manhattan apartment and weekends in the Hideaway. The other days were just hit-and-miss—sometimes they got together and other times they missed.

When alone at night, Helga often wondered loudly: Where were they heading? And what would eventually become of their marriage?

It was Wednesday again. Helga left the office early, and bought some flank steaks, a bottle of red wine, and some pastries and flowers to celebrate with Rob their weekly tryst in Manhattan.

Rob arrived slightly late because of the traffic.

As soon as he entered the apartment, they flung into each other's arms and kissed passionately. During a short pause, Rob said, "Helga, I have terrific news!"

"What? You surely look excited!"

"How do you like to celebrate our third anniversary in Moscow?"

"What?" And after a pause, "Great!"

"I have just been invited by the president of the Soviet Academy of Sciences to give the keynote speech at the meeting of the Federation of Engineering and Scientific Societies of the USSR in Moscow at the beginning of September and to receive an honorary membership of the Soviet Academy of Sciences."

Pulling out a telegram and waving it at Helga, he continued, "I will be the only American so honored as an honorary academician. And we can always go to Moscow a week ahead of the meeting and keep the time to ourselves. Can't we?"

With a short "mmm," she continued with their kiss.

The rest of the evening was properly spent on how to have their second honeymoon in Moscow.

They arrived via Paris on an Air France flight at the Sheremetyevo Airport about fifteen miles from Moscow one clear afternoon of the first Sunday in September, and were met by Professor Avenir A. Serov and Mr. Nikita Loginov of the Institute of Automatics and Telecommunications, which was a prestigious institute under the USSR Academy of Sciences and which had 600 scientific workers and twenty-five chair professors. Professor Serov, director of the Institute and a very distinguished scientist-engineer, officially greeted Dr. and Mrs. Jack Rob Tains.

"Dr. and Mrs. Tains, we are most honored by your visit. In behalf of President Keldysh of the Academy and all our co-workers, I wish to extend to you our most cordial welcome for your visit to Moscow," Professor Serov said with a broad smile.

"Professor Serov, the pleasure is entirely ours. I feel extremely honored by this invitation, and we look forward, with great enthusiasm, to visiting your country and my professional colleagues in the Soviet Union."

"Dr. and Mrs. Tains, allow me to introduce you to Mr. Nikita Loginov. Mr. Loginov is a scientific worker at the Institute and speaks very good English. He will be Dr. Tains' official guide during the meeting and his official visits to the various institutes. However, we also have instructions that you do not wish to be disturbed during the immediate week. Mr. Loginov will actually come to pick Dr. Tains up a week from Monday. Meanwhile, we have arranged a car with an English-speaking chauffeur at your disposal this week."

They small-talked on their way to the hotel.

They arrived at Hotel Russia (phonetically, Hotel Russiya) in half an hour. The hotel was 90% finished and only 2,000 of the total 6,000 rooms were being used. When completed, the hotel was to become the largest hotel in the world. The Tainses were given a lovely suite with windows, having a fantastic view of the Kremlin and the Moscow River. It literally faced the tower of the southeastern tip of the Kremlin wall and led to an oblique view of the Red Square.

How would you like to have a whole week of a carefree vacation in this fabulous and fascinating city of Moscow? The Tainses had it. And they did not waste any time to kick it off.

Following the tourist guide and its enclosed map which they obtained from their travel agent, they took off from their hotel on foot and went to the Aragvi Restaurant at 6 Ulitsa Gorky for Georgian food to celebrate their third anniversary.

After they were warmly greeted by the waiter, Rob ordered vodka, and then summoned up enough courage to use his Russian to order their dinner. He had learned some Russian from a couple of language books and phonograph records for tourists during the past several weeks.

"*Tsiplyata tabaka*," Rob ordered and pointed his index finger at Helga and himself, conveying the idea that it was two orders. *Tsiplyata tabaka* or chick on the bricks was the specialty of the house. The friendly waiter, wearing a big mustache, nodded his head approvingly.

The Aragvi was a cozy place with soft Georgian music and a warm character of its own. And everybody seemed to be joyfully indulged in his own pleasure—be it his vodka, food, company, fancy or fantasy, and was entirely oblivious of their neighbors at the next table. Rob and Helga just loved it.

Across their small table, Rob and Helga just held their hands like newlyweds. Then Helga squeezed Rob's hands with strong grips, and both of them ritualistically chimed in, "Love is not a comma, but a period. A helluva grip of a period!"

For most of the evening, they smiled at each other. They were too contented with the exotic food, drinks, music and atmosphere to say much. It was an exotic moment of their life, and they were intoxicated.

They were among the last to leave the restaurant, and they walked slowly. Unlike their hurried trip to the restaurant, their return walk had a heightened sense of consciousness and sensibility—they felt as if they had never known each other well until this moment. And the world was never as beautiful as it was today.

They passed gorgeous buildings and walked along beautiful boulevards. They had yet to discover their identities during the daytime in the days ahead.

Suddenly, they came up the northern end of the Red Square. They had seen the Red Square with massive military parades in magazines or on television before. But, now it was the real thing. As they turned there, it opened up and its expanse awed them. With the great red wall of the Kremlin on their right and a massive sandstone-colored building on their left (which, they discovered the next day, was the Gum Department Store), the Red Square was a colossal roadway leading to the St. Basil's Cathedral which was, in Rob's own words, the exotic castle with soft-ice-cream-cone domes.

"Helga, do you feel like Dorothy in *The Wizard of Oz* marching on to the castle?" Rob asked.

"Yes, I do. But, I don't know whether I am taking along a cowardly lion or a rusty tinman or a straw-stuffed scarecrow with me."

"I believe that the guy you take along has the combined attributes of all three."

Helga quipped, "You mean: cowardly, rusty and straw-stuffed?"

"No," Rob was quick to respond. "He is lion-hearted and tin-armored, and scares the hell out of scoundrels."

Helga winked. And Rob enveloped her with his strong arms. They embraced and kissed, and felt that the world was at their feet. Literally, it was so; the Red Square which had seen numerous historical events was now witnessing a great love.

They walked, stopped, and walked, marching toward the fantasy land of their dream, the incomparable St. Basil's Cathedral of magnificent towers.

For the following week, Rob and Helga were bustling with sightseeing, dining and theatregoing in Moscow.

As a ballet dancer during her grade and high school days in Tenafly, Helga particularly appreciated the theatres in Russia. Naturally, they went to see a ballet, *Swan Lake*, at the Bolshoi Theatre, the home of the classical 19th-century ballet and now the genuine classical Russian ballet. They also went to other theatres. Among the ballets and operas they saw were *Giselle*, *La Boheme*, *Madame Butterfly*, and *Asel*, with a revolutionary plot and zest.

What intrigued Rob and Helga most was the puppet show, "An Unusual Concert," directed by People's Artist of the U.S.S.R., S. V. Obraztsov, in the State Central Puppet Theatre. Thanks to a printed program in English and a Canadian gentleman who spoke fluent Russian and sat next to Helga and explained and translated some of the actions and conversations, they enjoyed this unique show complete with conductor, cellist, soprano, tango dancers, baritone, pianist, poet, Gypsy choir and lion-tamer.

When Madame Aroni and Monsieur Mac danced their MAC-ARONI (or macaroni) tango, it was hilarious. Mac kept on flipping Aroni almost off stage or carried her in his arm precariously, arousing laughter and gasps among the audience.

During the intermission that followed, Rob said, "Helga, I never did know that love could be that romantic!"

"You know you are not the type," she responded.

"Why?"

"Because——" After some hesitation, she collected her thoughts and said, "Because you don't like macaroni."

Rob seemed to be lost in words, but laughed. And she grimaced in triumph.

Mr. John Jones-Jonini, the baritone puppet that followed in the program, had a remarkable resemblance in his look to Rob's, except for his tousled hair and bushy eyebrows.

"Rob-Jonini, I did not know that you are an escapee from the Moscow puppet troupe," Helga whispered.

"Shsh——" Rob put his finger over his mouth. "Don't let them hear what you have said, or they might get me back."

"What's wrong? Life might be more interesting in the theatre."

"No, I am not going unless they create a Helga doll to be in my show."

"Nope, I am too fat to be a cute doll," Helga smirked. "Besides, the only song I can sing is 'Happy Birthday.' "

That night they walked back to the hotel in jump steps alternately tangoing, waltzing and gaiting like two happy puppets.

Rob and Helga claimed that they had never seen as many museums in one city as they saw in Moscow. And they were impressed.

Today, they were to visit the Kremlin. First they joined the line for foreign visitors (which was considerably shorter than the one for native visitors) to visit Lenin's tomb just outside the Kremlin wall, and reached the rose-granite mausoleum in thirty minutes. Inside, an atmosphere of dignity and reverence prevailed. Like others in the line, Rob and Helga walked in silence with awe and veneration, stepping down to the floodlit, glass-covered casket and briefly looking at the recumbent figure in dark suit, white shirt and black tie. And flash cards of memory from the history book loomed large before them: Lenin, The October Revolution, The Civil War. And when they emerged from the tomb, they began to recover from a profound emotional and moving experience.

After they entered the Kremlin walls, they were met by their intourist guide and were pleasantly surprised. A stunning blonde in white blouse, black skirt and jacket, and carrying a red handbag, she spoke to them in perfect English, "How do you do?" Then they began with their tour.

They toured around the sixty-five-acre stockade of palaces, cathedrals and office buildings which are collectively called the Kremlin, whose name has inspired so much veneration and fear. They also stopped at the giant Czar Bell and Ivan the Great Bell Tower where they posed for pictures.

Then they came upon the magnificent cathedrals: The Apostles Cathedral, The Annunciation or Blagovoshchensky Cathedral where the czars were christened and wed, The Assumption or Uspensky Cathedral where they were crowned, and Archangel or Arkhangelsky Cathedral where many of them were buried.

When they were in the Arkhangelsky Cathedral, the guide explained, "When this cathedral was completed, Ivan the Terrible, who built this magnificent structure, ordered the architect's eyes gouged so that he could never design another structure to match the beauty of this one."

"Watch out, Rob," Helga jested "for what they may do to you when you finish your 'space' project."

"I wouldn't mind at all if, after 400 years, there are willowy guides who will tell tourists about my accomplishments, like what our guide did today about the architect."

This was the last of their sightseeing days, as the meeting of the Federation of Engineering and Scientific Societies (FESS) of the USSR was going to open tomorrow.

They went by car to the Novodevichy Convent on the East Bank of the Moscow River. An ensemble of historical and architectural monuments of the 16th and 17th centuries, it was the best preserved of all monasteries in Moscow. Like the St. Basil's Cathedral, it also had soft-ice-cream-cone domes.

The large cathedral was now a museum, but in the small church, there were still regular religious services.

Rob and Helga were fortunate to witness the religious service. Hand in hand, they stood in the back and observed the proceedings. They too felt consecrated with a new blessing for their marriage. And they felt very much at home.

After the service, they went to the cemetery area of the Convent where many well-known Russian artists, composers, writers and other notables were buried side by side. They also watched the Moscow River and its verdant banks.

After they left the Convent they stopped at a nearby dollar shop where you could buy luxurious Russian products often at bargain prices and where the Russian government earned hard currencies from its tourists through these bargains.

Rob bought an amber necklace pendant and a sable cape. After he put the pendant around Helga's neck and the cape over her shoulders, he kissed her in front of the store counter and said, "You are beautiful."

The sales girls were stunned, but Helga was secretly very happy.

Six

The annual meeting of the Federation of Engineering and Scientific Societies (FESS) of the USSR began today and had its keynote session scheduled at 2 P.M. in the 6,000-seat Conference Hall of the Palace of Congresses under the watchful eyes of Lenin, whose likeness adorned the giant curtain.

Rob sat nervously on the stage along with other dignitaries. Dr. Nicholas Krylov, president of the FESS, delivered the welcoming address and expressed his distinct pleasure to have a most distinguished American engineer-scientist, Dr. Jack Rob Tains, as their keynote speaker. And a glowing introduction of Dr. Tains, delivered by Professor Avenir A. Serov, director of the Institute of Automatic and Telecommunications, followed, with this concluding statement: "Today, we have with us the creator of the 'robot,' which is also known as the 'Human Intelligence Model' and which is inconceivably used as the workhorse in almost all electronic, communication, computer, and control systems now in use throughout the world. And his invention-and-theory and his present responsibility may yet usher in a new era for all of us, an era of space exploration with human moon landings merely as the first step."

The audience listened in silence and awe. So did Rob on the stage and Helga sitting among the audience.

After a thunderous ovation, Rob rose from the seat and walked to the rostrum. He thanked his Russian hosts profusely, "I am grateful for this invitation and for this opportunity to meet with you. Professor Serov was very kind to give me a great deal of credit for my accomplishments. But, the frontiers of science and technology have really been established collectively by scientists and engineers throughout the world. And let me use this opportunity to thank all of you and your professional colleagues throughout the Soviet Union for your outstanding contributions." And another round of applause ensued.

Now, Rob proceeded with his keynote speech on "The Robot and Its Robotosyncrasies." He had been forewarned that, since the audience was to be a heterogeneous group of engineers and scientists of many different disciplines, the speech should be of popular nature and cater to the laymen rather than the specialists. An abbreviated version of Rob's speech[1] is given with captions and in first-person expressions as follows (The reader may skip this speech *or* read a simpler version in chapter 2 of Part I, and continue on):

The Crude Robot

For a *command* or *input* x in terms of a signal, we want a robot to have a *response* or *effect* y (which is often called a "*control variable*" in technical terms) to obey and follow the command or input.

First let's consider a "robot without intelligence" or a "crude robot" as shown in Fig. 1 of the slide projection on the screen (which is also Fig. 1 of this book). A crude robot, often called a *controller*, is designed with many system components (e.g., transistors, diodes, resistors or capacitors if the crude robot is an amplifier; amplifiers, motors, networks and/or other components if the crude robot is a certain type of control system) to do a specific job. It has a *gain* G relating the output y to input x with the relation:

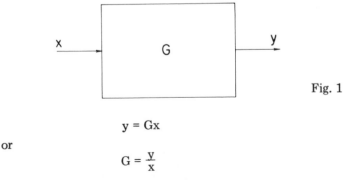

Fig. 1

$$y = Gx$$

or

$$G = \frac{y}{x}$$

The Bandwidth: The Breadth of a Robot's Mind

The input to a robot is usually a signal.

We are all familiar what is a signal. When we hit a piano key, we have an *elementary signal* of a certain frequency f.

As an example, we may have elementary signals of these frequencies:

f_1 = 10 cps (or 10 hertz)
f_2 = 35 cps
f_3 = 10,000 cps
f_4 = 1,000,000 cps

The *frequency* f of an elementary signal is the number of vibrations (or cycles) per second with a unit in cps (or cycles per second) or hertz (meaning cps).

An elementary signal has a *frequency* f in cps as well as a *magnitude* x (or y), say, in volts or other units.

For a controller (i.e., a crude robot) or a robot (which has yet to be introduced), we may feed in an input x at a certain frequency f, measure its output y, and compute its ratio or gain G=y/x at this frequency. By repeating this step for different frenquencies, we may obtain the values of G at different frequencies and may plot a *gain characteristic* G as shown in Fig. 2*b*.

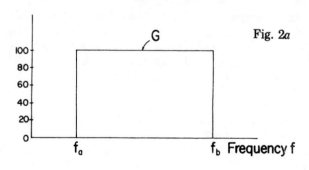

Fig. 2*a*

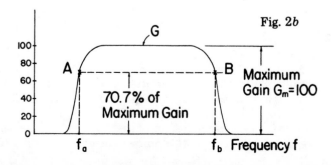

Fig. 2*b*

As can be shown mathematically, when the gain characteristic G of a controller (or robot) is a *constant*, say, within certain reasonable frequency limits f_a and f_b as shown in Fig. 2*a*, the output y is almost identical in form to the input x. This is the most desirable situation, as the output or response y follows the input x faithfully just as the controller (or robot) is required to do.

Note that, for any frequency f between f_a and f_b in the example of Fig. 2*a*, if the input is x = 1, the output is y = Gx = 100. On the other hand, for any frequency outside these limits f_a and f_b, e.g., $f < f_a$ (or $f > f_b$), if the input is x = 1, the output is y = Gx = 0. It is easy to define the *bandwidth* of the controller or robot of this "ideal case" with a constant gain to be:

$$W = f_b - f_a$$

For humans, we have our range of emotion, and for robots, they have their range of frequency. The bandwidth is, therefore, the breadth of a robot's mind to accommodate signals or commands.

However, the gain characteristic G in Fig. 2*a* with sharp cutoff frequencies f_a and f_b for its passband is an *ideal* case, and exists only mathematically.

In the physical world, the gain characteristic G is a curve and is therefore called a function of the frequency f (namely, the value of G is different at different frequencies), as is shown in Fig. 2*b* and note that it tapers off at two ends gradually rather than abruptly. Then, how do we define its bandwidth? In these physical cases, we define f_a and f_b to be the frequencies at which the gain G drops to 70.7% of the maximum gain (and f_a and f_b are called "half-power frequencies," at which the output y has ½ of the maximum power and has $1/\sqrt{2} = 0.707$ or 70.7% of the maximum magnitude). The bandwidth, as the breadth of the robot's mind, is still defined to be:

$$W = f_b - f_a$$

The Robot As a Human Intelligence Model (H.I.M.)

Many years ago, an idea struck me when I tried to park my automobile in a space along the curb of a busy street in a Long Island township. At each moment, I compared my "objective," i.e., the final desired parked position of the car, with my "observation" of the current status, i.e., the current position of the car. In this process the *actual input* w to my mind was therefore the difference (w = x-y) of the *reference input* x (namely, my objective) and *observed output* y (namely, my observation); and this process is a continuous one. This is the concept of my Human Intelligence Model (H.I.M.). I don't think that the human being is a supreme being with an innate complex mental mechanism. Rather, it is the simple concept of the H.I.M., applied repetitively and continuously, that is responsible for many human qualities and wisdom.

Then, why can't we apply the concept of the H.I.M. to the crude robot, making it a *robot* with human intelligence? This is how I introduced the system model for a robot in Fig. 3*a*. The circular device in the schematic is a sensing device or comparator, taking two inputs x and z and offering its difference w = x-z as output. Note here in Fig. 3*a* that, for a reference input x and an observed output y, their difference w = x-z = x-y is fed into the controller for processing.

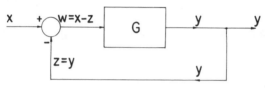

Fig. 3*a*

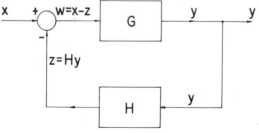

Fig. 3*b*

Later, I introduced an additional degree of freedom in the system model of robot in Fig. 3*b* when a "crude robot" with a gain H is inserted in the feedback loop. The simpler model in Fig. 3*a* is, therefore, its special case with H = 1.

For the system model in Fig. 3*b* we want to obtain its input-output relation, namely, an equation relating x to y.

From Fig. 3b, we have these relations:

$$z = Hy \qquad (1)$$
$$w = x\text{-}z \qquad (2)$$
$$y = Gw \qquad (3)$$

From these three simultaneous equations, we can readily eliminate two variables w and z, and obtain Eq. (A) below relating x to y.

For completeness, we shall carry out this process of eliminating w and z [which the reader may *omit* and proceed to Eq. (A) directly.]

First substituting (1) into (2), we obtain:

$$w = x - Hy \qquad (4)$$

We then substitute (4) into (3):

$$y = G\,(x\text{-}Hy) = Gx - GHy$$

which may be written in the form of:
$$y + GHy = Gx$$

or

$$y\,(1+GH) = Gx \qquad (5)$$

Changing the form of the above, we have the *fundamental equation* of the robot:

$$y = \frac{G}{1+GH}\ x \qquad (A)$$

Defining the *system gain* $G^* = y/x$ as in Fig. 4, the fundamental equation may also be written in this form:

$$G^* = \frac{G}{1+GH} \qquad (B)$$

The fundamental equation (A) or (B) will be found to be a powerful tool to establish the various robotosyncrasies, as we shall do in the remaining portion of this lecture.

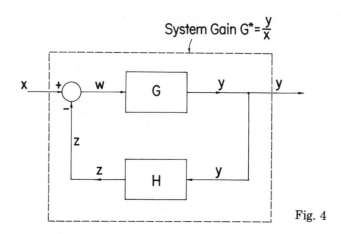

Fig. 4

Robotosyncrasy 1: Broadminded

We can readily show that a robot is much more broad-minded than the crude robot or controller.

We shall postpone our illustrative example to establish both idiosyncrasies 1 and 2 together.

Robotosyncrasy 2: Even-Tempered

We shall also show that a robot is much more even-tempered than the crude robot or controller.

Illustrative Example

Let's consider the system model in Fig. 3b with H = 0.09 (which is typical for a feedback-amplifier network arrangement).

Let's assume the gain characteristic G of the controller be represented by this table:

FREQUENCY (in cps)		GAIN G
f_1 =	10	15
f_2 =	100	35
f_3 =	1,000	55
f_4 =	10,000	100
f_5 =	100,000	20
f_6 =	1,000,000	10

TABLE A

or by the graph in Fig. 5a.

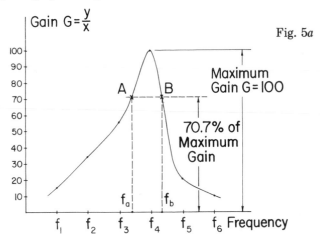

Fig. 5a

(Note that, for f_1 = 10; f_2 = 100; f_3 = 1,000; . . ., the points are equal-spaced along the frequency axis in Fig. 5a. This is because the frequency axis is on a *logarithmic scale*. It is a standard practice to use a specially prepared graph paper with a "logarithmic scale" for the frequency axis in plotting gain characteristics.)

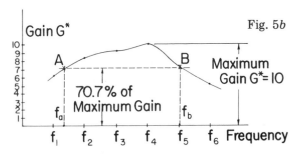

Fig. 5b

Note that with the gain characteristic in Table A or Fig. 5a for a constant input x = 1, the output will be y = 35 at the frequency f_2 and y = 100 at the frequency f_4. The outputs y = 35 and y = 100 are radically different for the same input x = 1. The controller or crude robot is very uneven-tempered and is temperamental in the sense that it will not obey commands uniformly at different frequencies. In other words, it is a disreputable (crude) robot.

In general, we choose the controller or crude robot with a gain characteristic much better and more even-tempered than the one in Fig. 5a. However, we deliberately choose a bad one for illustration.

From Fig. 5a (where f_1, f_2 . . .are defined in Table A), we may readily determine:

$$f_a = 3{,}000 \text{ cps}$$
$$f_b = 30{,}000 \text{ cps}$$

and the bandwidth of the controller or crude robot is, therefore,

$$W = f_b - f_a = 27{,}000 \text{ cps}$$

which is a rather narrow bandwidth. And the controller or crude robot is therefore rather "narrow-minded."

We shall now investigate the robot with the system model in Fig. 3b and H = 0.90. Using the values of G in Table A in the fundamental equation (B) we may compute, for example at $f_1 = 10$,

$$G^* = \frac{15}{1+15 \times 0.09}$$

$$= \frac{15}{1+1.35} = \frac{15}{2.35} = 6.4 \qquad \text{(for } f_1 = 10)$$

Repeating for other frequencies f_2, f_3 . . .we now have

FREQUENCY (in cps)		GAIN G^*
$f_1 =$	10	6.4
$f_2 =$	100	8.4
$f_3 =$	1,000	9.3
$f_4 =$	10,000	10
$f_5 =$	100,000	7.2
$f_6 =$	1,000,000	5.0

TABLE B

And these data are now plotted in Fig. 5b.

From Fig. 5b (where f_1, f_2. . . are defined in Table B), we may readily determine:

$$f_a = \quad 20 \text{ cps}$$
$$f_b = \quad 100{,}000 \text{ cps}$$

and the bandwidth of the robot with the system model in Fig. 3*b* is now:

$$W = f_b - f_a = 100{,}000 - 20 = 99{,}980 \text{ cps}$$

which is much more *broad-minded* than that (27,000 cps) of the controller or crude robot.

By comparing Figs. 5*b* and 5*a* it is easy to see that the robt is *more even-tempered* than the controller or crude robot.

Robotosyncrasy 3: Agile

Assuming a step-function input x as command, the "ideal" output y as effect or response is exactly the *same* as the input in form, as is shown in Fig. 6. However, this is an impossibility for a physical system.

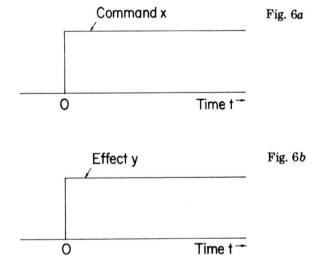

Command x   Fig. 6*a*

O   Time t →

Effect y   Fig. 6*b*

O   Time t →

Rob then proceeded to derive with mathematical techniques (as shown in Appendix A of this PART) that for an

input x as shown in Fig. 7a, its output y is shown in Fig. 7b
with a small time delay t_d and a buildup time t_B, where

$$t_B = \frac{1}{2f_c} = \frac{1}{W}$$

and W is the bandwidth (and f_c is the cutoff frequency of a
low-pass controller or robot).

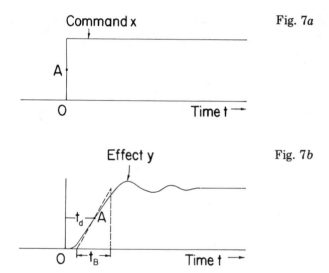

Command x   Fig. 7a

Effect y   Fig. 7b

It is now easy to see that for zero time delay (i.e., t_d=0)
and an infinitely large bandwidth (i.e., W=∞, and t_B=0), the
output y in Fig. 7b will be almost *identical* to the input x in
Fig. 7a in form.

Ignoring the small delay t_d, we note that the broader the
bandwidth W, the smaller the buildup time t_B and the more
agile is the crude robot or the robot in responding to the
command.

In the above *Illustrative Example*, the controller or crude
robot has a bandwidth W = 27,000 or a buildup time t_B =
1/W = 37 x 10^{-6} secs; and the robot has a bandwidth of W =
99,980 or a buildup time t_B = 1/W = 10 x 10^{-6} secs. We can
readily note that the robot is much *more agile* (3.7 times faster
for this case) than the crude robot in responding to the
command.

Robotosyncrasy 4: Stable and Exercising Moderation

Let's now consider the system model in Fig. 8a in which the reference input x is directly compared to the observed output y, where their difference or "error" w = x-y is fed into the process. The reference input or command x is assumed to be a step function as shown in Fig. 8b.

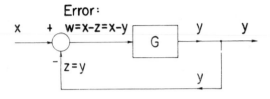

Fig. 8a

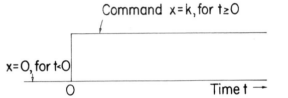

Fig. 8b

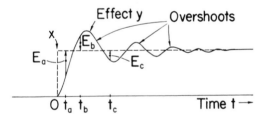

Fig. 8c

A "robot" system operation could be stable or unstable. There are criteria for detecting stability which I shall also discuss in this lecture. Using these stability criteria and compensation design techniques, "robot" systems are always designed to be stable.

Now, let's briefly discuss stability vs. instability.

Stable Operation

In Fig. 8c we sketch the input or command x in dotted form, as a reference for comparison with the output or effect y. And we note:

1. At the instant $t = t_a$, x is larger than y in magnitude, and their difference:

$$x - y = E_a$$

is positive. In Fig. 8c we use an "upward" arrowhead to indicate E_a as a positive quantity. As E_a is *positive*, the controller tends to *increase* or boost the output magnitude y.

2. As the output magnitude y increases, y becomes greater than x.

3. Now at the instant $t = t_b$, y is greater than x, and their difference

$$x - y = E_b$$

is negative. In Fig. 8c, we use an "downward" arrowhead to indicate E_b is a negative quantity. As E_b is *negative*, the controller tends to *decrease* or pull back the output y.

4. As the output magnitude y decreases, y gradually becomes smaller than x.

5. Now at the instant $t = t_c$, y is smaller than x and their difference

$$x - y = E_c$$

is positive again. As E_c is *positive*, the controller tends to *increase* or boost the output y.

6. Repeat this process many times until the output y settles down to a constant value as the input x. Here we have the output y (solid curve) as shown in Fig. 8c, which is a reasonable approximation of the input x in Fig. 8b.

In this operation, the controller *reacts* with positive or negative error (i.e., E_a, E_b, E_c ...), and *compensates* the output y by increasing or decreasing it.

We are very fortunate that a well-designed robot neither *overreacts* nor *overcompensates*. As a result, we have stable operations, as just described, with decreasing overshoots and with the output or effect y settling to a constant value as prescribed.

Unstable Operation

On the other hand, an unstable operation is described in

Fig. 8*d*. The above detailed steps describing a stable operation may be used to describe an unstable operation with a slight modification: the robot *overreacts* and *overcompensates* with the result that the overshoots grow with each cycle as shown in Fig. 8*d*.

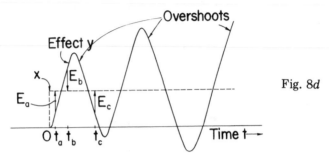

Fig. 8*d*

With the growing overshoots, where does it lead to?

Theoretically (namely, mathematically), it will lead to overshoots of infinite magnitude, and the overshoots will perpetually grow in magnitude. In technical terms, the output or effect y is called a *runaway transient* response.

The principles of physical science dictate that a pheno-

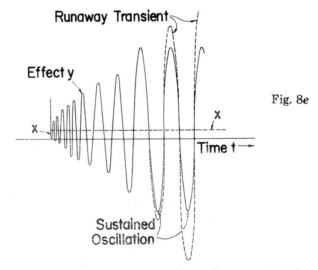

Fig. 8*e*

menon (say, an output response) of an infinite magnitude requires an infinite amount of energy. And no creature or device or system has an infinite amount of energy, not even a rich robot. Therefore, the unstable response y in Fig. 8*d* cannot grow indefinitely. Instead, it will grow into a *sustained oscillation*, limited by the energy content of the robot, as shown in Fig. 8e (where Fig. 8d has been shrunk in scale and extended).

Now, let's see what happens with an unstable operation. For a given step-function input or command x as in Fig. 8b we wish to obtain an output or effect y identical to x in form. Instead, we obtain the runaway transient y in Fig. 8d or the sustained oscillation y in Fig. 8e. Of course these responses are not what we intend to have. The robot is functioning unintelligently in a retarded fashion, or is running wild or mad in an insane mood.

In the human world, we put the retarded and the insane in institutions or asylums, and they do live a useful life in a perhaps limited sense. What about the robots? Fortunately, the oscillatory robots are useful, in a more limited sense than the (respectable) robots, as *oscillators*. For the use of oscillators, particularly, in electronic circuits or systems, ask any electrical engineer.

Remarks

It is gratifying to note that even a robot believes in and exercises *moderation*—not to overreact, and not to over-compensate. It is "moderation" that is responsible for the successful operation of the robot in the stable mood.

However, this "moderation" does not come by accident. It is carefully designed by men of ingenuity. And I shall later discuss these design considerations under the captions of stability criteria and compensation techniques.

Robotosyncracy 5: Visionary

As human beings, our mental process is actuated by signals in terms of intelligence or emotion. A signal is often accompanied by a noise (i.e., disturbance or doubt) and is corrupted by this noise. If the noise is too large, the signal becomes non-decipherable and meaningless. This is also true for the robot.

We shall consider the signal-and-noise relations of both 1) a controller or crude robot and 2) a robot, where the noise comes from the internal sources, and compare these two situations.

In order to obtain quantitative measures, we shall define the *signal/noise ratio* as

$$\frac{\text{Signal}}{\text{Noise}} = \frac{x}{n} \left(\text{or} = \frac{y}{m}\right)$$

where a signal x (or y) is accomplished by a noise n (or m) and where x and n (or y and m) are magnitudes in the same unit (e.g., volt). In general, we shall find that *the higher the signal-noise ratio, the more desirable is the situation* in the sense that the signal is less corrupted by the noise.

First, let's consider the controller or crude robot with a gain G. In order to accommodate an internal noise source n, we shall use the schematic in Fig. 9a where the controller is split into two components with gains G_1 and G_2, and G = $G_1 G_2$. The output of G_1 and the noise n add together, forming the input to G_2.

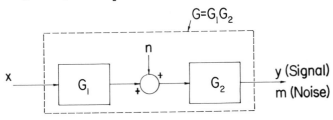

Fig. 9a

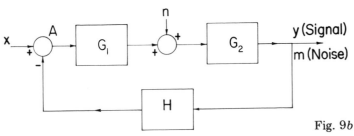

Fig. 9b

We then consider the robot with an internal noise source n and a system model as shown in Fig. 9b.

Assuming some typical operating values of the various parameters and holding the "same" noise input n for both cases, Rob was able to obtain a) a signal/noise ratio, y/m = 10, for the output of the controller or crude robot in Fig. 9a and b) a signal/noise ratio, y/m = 100, for the output of the robot in Fig. 9b.

It is now obvious that the robot, with a much higher signal/noise in its output, has its output or effect much less corrupted by its internal noise in terms of internal disturbances and self-doubt. In other words, the robot is more *visionary* than a controller or crude robot.

Robotosyncracy 6: Resolute and Reliable

A controller or crude robot will have dozens, hundreds, thousands (or even millions) of components. To operate satisfactorily, it must have a constant gain G, say G = 500, for its operating frequency range (and will allow the gain characteristic to drop outside this range). Suppose that one or several of these dozens or hundreds or thousands of components "age" or become "faulty," the gain could drop to G = 450 and the result could be catastrophic. (Consider this controller is in the control system of a spaceship!) But, to make all these dozens, hundreds or thousands of components "precision" and "fault-proof" components is both enormously costly and almost impossible. Then, what shall we do?

The robot with its system model actually saved us. This is perhaps the most important of all robotosyncrasies and is based upon a very simple principle.

Let's consider the fundamental equation (B), which has been earlier derived for the system model in Fig. 4:

$$G^* = \frac{G}{1+GH}$$

Suppose that we divide both the numerator and denominator on the right side of the equation by G, thus obtaining:

$$G^* = \frac{G/G}{(1+GH)/G} \qquad \text{or} \qquad G^* = \frac{1}{(1/G)+H} \qquad (C)$$

When $1/G$ is very small (and it does matter whether G = 450 or G = 500) as compared to H, Equation (C) reduces to the following approximation:

$$G^* \cong \frac{1}{H} \qquad (D)$$

Even when several components in the controller failed and its gain dropped from G = 500 to G = 450, the system gain G^* is unchanged and is dependent upon the value of H alone. And we need only to make H reliable to have a reliable robot system.

For some systems, e.g., transistor feedback amplifiers, the H box in Fig. 4 is merely a single resistor. By using a single expensive reliable "precision" resistor for the H box (and using numerous cheap components for the G box, i.e., the controller or transistor amplifiers), a reliable system is created.

We have now shown that a robot is far more *resolute* and *reliable*, with amazing results, than a controller or crude robot.

Criteria for Stability

Rob then went on to discuss the system function, characteristic equation, and open-loop transfer function in terms of stability (Chen[1], pp. 474-478), as well as the Routh's Criterion for Stability (loc. cit., pp. 478-485), Nyquit Criterion for Stability (loc. cit., pp. 489-514), and Root-Locus Method (loc. cit., pp. 487-489).

He also remarked that we must obtain the system function, characteristic equation or open-loop transfer function before we can apply any of these criteria for stability, and that we may use singal-flow techniques (loc. cit., pp. 387-465), to obtain these funtions or equations.

[1] W.H. Chen, "The Analysis of Linear Systems," McGraw-Hill, 1963 *or* other appropriate books on closed-loop control systems or linear systems.

Compensation for Improving Stability and Performance

With the criteria for determining stability now at our disposal, Rob continued his discussion with the methods of compensation for improving stability and performance for robot design (loc. cit., pp. 549-569).

Concluding Remarks

Ladies and gentlemen, I have just concluded my technical presentation. I do, however, have some general concluding remarks to make.

Again, I wish to thank you for this significant honor of inviting me to present this keynote speech.

Our two nations, the two superpowers of our time, have been at odds on many issues. However, I am extremely pleased that we now have a common language, the Robotese, that has brought us together today.

We have now learned that the robot is *broad-minded, even-tempered, agile, stable, visionary, resolute* and *reliable*.

Shall we then learn from him, the robot of our own creation, these virtues?

We have also learned that a wise robot resists self-doubt, exercises moderation, and never overreacts. And now shall we follow his example to rediscover these cardinal rules of nature for the salvation of our own world?"

When a standing ovation ebbed, President M. V. Keldysh of the Academy of Sciences of the USSR came forward to the rostrum, presiding over the presentation ceremony. Professor Serov read Dr. Jack Rob Tains' biography and presented Rob to President Keldysh for the conferring of a honorary membership of the Soviet Academy of Sciences with the citation "For extraordinary contributions and services to mankind."

Academician Rob Tains took a deep bow as the finale of this historical event.

Seven

Except for the keynote session on Monday which Helga attended as the speaker and honoree's wife, all other sessions for the annual meeting of the Federation of Engineering and Scientific Societies of the USSR for the rest of the week were only open to registered participants. As Rob was busy attending these sessions, Helga had all the time in the world to do whatever she pleased.

Tuesday, after a morning stroll to the Gum Department Store and the Detsky Mir (the Children's Department Store), where she bought some small souvenirs for friends and her niece Charmaine and nephew Michael, she returned to Hotel Russia for lunch.

It was a Russian custom at that time to seat guests to near capacity of each table in the hotel restaurant. And Helga was seated at a table of two with a beautiful girl with auburn hair and a pair of oversized gold-rimmed, slightly tinted glasses.

"Hi," said Helga, fully expecting the other girl to be an American.

The other girl flashed a smile, extended her hand to Helga, and said, "I am Jane Green from New York."

They shook hands. And Helga now took her seat.

Helga suddenly began to realize Jane's familiar look. *Did I see her on the LIFE magazine's cover? Yes, I did. Now, even her name is also right.*

"You are——" Helga stumbled. "Oh, you are the author of *The Sex Object?*"

"Yes. Have you read the book?

"Yes, I have," Helga replied. "And I really enjoyed it. By the way, my name is Helga Tains and I am also from New York."

A friendship began to blossom over an American-style lunch in a Moscow hotel. And they chatted away over many cups of black coffee.

Helga and Jane spent the next three and a half days together. They visited the Pushkin Fine Arts Museum, Lenin Library, Tchaikovsky Conservatory and Tretyakov Picture Gallery. They rode on the Metro (subway) and admired the elegant design of its stations. They toured the USSR Exhibition of Economic Achievement which, Helga swore, was fifteen parks in one, and they had to ride Disneyland-style trams to tour around. There they discovered the only Coke and hotdog stand they saw in all Moscow, and naturally they stopped for a snack. And they even saw the Circus on Ice, one of the top attractions in Moscow.

Their conversation was anything but on culture, arts and music, even though they were in museum, library and conservatory. It was almost a monologue by Jane, interspersed by Helga's mms, yeahs and ohs. Samples:

"Helga, men and women were not created equal. Even since the beginning in the days of Eve, women have been the underdog subject to the abuse, prejudice, ridicule, whim, wishes and bestial desires of men."

"Mmm," Helga uttered the sound almost inaudibly.

"Just think about the medieval chastity belt. It symbolizes man's age-old attitude toward woman. She was his property, to use and misuse, and to control and dominate."

"Yeah."

"But, Helga, the modern man is no better. As a matter of fact, the average man, including the so-called liberal, wants a passive sex object *cum* house maid *cum* baby nurse, and bosses her around. Perhaps all men are male chauvinists on some level."

"Oh," Helga murmured. She was interested but didn't know how to take an active part in the conversation.

On Friday, they took a cruise on the Moscow River. As the pleasure boat moved along its generally westward course, the ever-emerging landscape along the banks presented itself and was just lovely.

Unlike the preceding days, Jane was not very talkative. She was not pushy at all.

"Jane, are you married?" Helga asked.

"I considered myself *almost married*."

"Are you going to marry him?"

"Why should I?" Jane winked.

And there was a silence. A dead silence for a while.

Then Jane continued, "Marriage restricts your freedom and makes you half of a legal entity—sort of half a person."

"Is this your reason to stay unmarried—pardon me, almost married?" Helga was curious.

"Helga, I just cannot afford to get married. As women, we also have our cause to fight for."

The cruiser moved gracefully along the river, passing by many familiar landmarks of the city: the Kremlin with its walls and cathedrals, the Gorky Recreation Park, the ski jump, the Moscow State University with its sprawling gigantic building complex like modern castles in cascade, the big bowl of the Lenin Central Stadium, and now the Novodevichy Convent.

All of a sudden Helga was seized by a choking nostalgic sensation. It was only five days ago, yes, five days ago, that she and Rob were standing hand in hand in one of these soft-ice-cream-cone-domed cathedrals of this convent, witnessing the religious proceedings and receiving blessings for their own wedded status. Now, as the cruiser moved along, she was leaving this scene behind her and cruising alone without Rob into new territories.

A montage of mental images flashed in her mind: the celebration of their third anniversary at the Aragvi, their romantic moments at the Red Square, the Bolshoi Theatre and the *Swan Lake*, Rob-Jonini and the puppet show, the Kremlin tour, the Convent, and Rob's exhibitionist's kiss in the Dollar Shop. While Helga was still wallowing in the warmth of her recent sweet memories, Rob was mysteriously fading away. His image became smaller and smaller in Helga's mental picture until it disappeared. She really felt that she wanted to cry.

For the rest of the journey, her mind was blank and she looked intoxicated. Nevertheless, she responded to Jane's conversation with monosyllabic yeahs. But she was not really listening.

When the cruiser docked again, Helga sort of woke up. This was the last afternoon of their spending time together in Moscow, and Rob was to join Helga this evening. When they returned to the hotel, they bade each other good-bye.

"Helga, let's get together in New York."

"O.K." Helga forced a smile on her face and nodded her head.

Rob and Helga left Moscow on a direct flight to New York. When dinner was served on the plane, Helga asked, "Rob, do you think that, ever since the beginning of civilization women have been the underdog of our society, subject to all kinds of abuse and indignation?"

Amused by the question, Rob shrugged. "You mean they have been well protected."

Rob dozed off for much of the journey. But Helga was very much awake, thinking loud.

Was Jane Green right for all what she had said? And her statements kept on haunting her during the flight. *"The average man wants a passive sex object . . . .Perhaps all men are male chauvinists on some level . . . .Marriage restricts your freedom and makes you half of a legal entity—sort of half a person . . . .As women, we have our cause to fight for."*

Helga was not in a position to sanctify all these statements. But, she saw plenty of ground to raise doubts and questions about our social institution, and even about her own life.

She remembered that, even before their trip to Russia, she was already wondering about their marital status: *Where were they heading? And what would eventually become of their marriage?*

Now, she indulged in some soul-searching. With their return to New York, were they also returning to an inevitable destiny—which she had feared? What was the cause of it? Their own busy work-schedule conflict? Or was it the social institution?

In a basically unequal society, was it her own guilt of not having served her master, the husband, well in the traditional sense that caused her anxiety and fear for the worse? Or was it the other extreme, that she felt guilty for her failure to "fight for our cause" as a woman, as called for by Jane Green? She was confused.

Although her love for Rob was absolute, she had some doubt about Rob's attitude about the sexes. Was he an average man and male chauvinist as Jane Green had referred to?

Her mind then roamed to her reporter days with the *San Francisco Examiner*. She was assigned to cover the collective-bargaining negotiations at a university. It had been her concept that the faculty and the administration consisting of deans and department heads should stand together and work together. But, no more. They were pitched against each other as adversaries. And it was this concept of adversaries, according to its advocates, that was responsible for social progress. And at that university, this concept prevailed with the unionization of its faculty, and the faculty and the administration continued to confront each other as adversaries.

Now, must men and women confront each other as adversaries? And must she and Rob confront each other as adversaries? Helga couldn't bear to think of the latter.

Helga was totally confused. But with Rob shrugging off even the simple question about the inequality of sexes and with Rob as a potential adversary, she must grapple with the problem all by herself.

Their plane landed at the J. F. Kennedy International Airport. It was a Sunday. They rented a car and drove to their Long Island Hideaway.

On their way home in the car, the sky was growling and it was exceedingly windy with thunder and lightning. But it was not raining.

That night a terrific storm seemed to hit the area. The rain was driving down hard with lightning flashing and thunder rumbling on all sides of their house. There was no letup throughout the night.

But by daybreak, the storm suddenly disappeared. Helga was still asleep. Rob got up and took a walk on his favorite beach.

At a distance, Rob noticed a white object on the beach near his five rocks, the five continents of his make-believe world during his childhood days. When he walked close, it was a dead sea gull. Apparently, it was a victim of the vicious stom last night. Rob stood there in silence for a while, paying his last respect. Then he removed the dead bird to the bluff and buried it there.

Eight

It had been a month since they returned from Moscow. And during this period, Helga visited Jane many times, and they had lunch together whenever they both were able to make it.

Tonight, Helga was to attend the Manhattan Chapter meeting of OWL, Organization for Women's Liberation, of which Jane was the founder, national chairperson, and speaker for this chapter for this monthly meeting.

After a brief business meeting for the chapter, the program chairperson introduced Jane as the speaker and capped her introduction by referring to Jane as "The OWL with vision and clout." And enthusiastic applause followed.

"Dear sisters," Jane began, and she was very direct with her speech and continued, "we are living in a man's world. Women are helpless because men control the mechanism of the society at all levels. In the government, we have yet to see a female President or a female justice of the Supreme Court. In the family, patriarchy has been practiced ever since the beginning of history on all continents. In business and in labor, women are at the lowest rungs of the organizational ladders."

After citing additional examples of injustice, Jane concluded, "It is no accident and it is a conspiracy of the first order—a conspiracy by *men*."

Raising herself to her full height and her voice to a state of indignation, she continued:

"Even marriage is a form of slavery and fraud. It is a combination of private burlesque, legalized rape, slave labor, forced submission, fraudulent contract, and curtailment of a woman's freedom without any assurance of love from a man. By legal means, we are put in a maximum security prison."

Jane was thunderously applauded. Helga was immensely impressed—not just by Jane's fiery speech, but also by the spirit and enthusiasm of 500 sisters gathered for this occasion.

Jane continued to rant against marriage as an institution and the deception by men, and then switched her target: "However, to blame the unilateral oppression of women by men as the sole cause for the social injustice is both misleading and dishonest. It is important that women should see that we have *connived* in this situation and that they have consented to become wall flowers, dress mannequins, super chattels, and passive sex objects to please 'our masters.' "

Jane now concluded her address:

"Sisters, do we realize for what purpose are we organized? Do we fully understand that we aim at nothing less than the dissolution of the male-dominated social structure? To achieve our goals, we must act—and act resolutely."

Amid roars of applause in a standing ovation, Jane retired from the platform. Helga was spellbound by Jane's magic. So was the audience.

For days, Helga remained mesmerized by Jane's oratory and kept asking herself: "Have I *connived* in this situation? And how can I act resolutely?"

It was Wednesday again, the evening of their weekly tryst in their Manhattan apartment.

Rob first went to the Chun Cha Fu Chinese Restaurant to pick up his order of Peking duck which he had secretly ordered last week. It was Helga's favorite. And his too. What a surprise Helga was to have!

As soon as Rob unlocked the apartment, he called, "Helga! Helga! Where are you?"

But, there was no answer.

Rob searched around. There was no trace of her in the entire apartment. But, there was a beautifully made-out flower arrangement on the dining table. Next to the vase, there was a letter with the envelope addressed in big letters: ROB

Rob immediately opened the envelope.

Dear Rob:

I must disappear from your life. As human beings, we all have our own missions.

Please do not try to find me.
I love you. I love you. And I will always love you in
my heart.

Yours,
Helga

Rob just couldn't believe it. He sat in the sofa with the
letter in his hand for hours. It was a great shock. And this
shock and unbelievability lingered on. For days. For weeks.

Nine

Rob was now back to his old working habit during his bachelor days. He literally "lived" in his office and laboratories on Long Island, days, evenings and weekends, partly because Helga was no longer a part of his life, and partly because of the work pressure for the Apollo projects to keep on schedule.

There were many companies engaged as contractors or subcontractors for Apollo support work, and Rob's CSI Aerospace Corporation wasn't even the largest. However, it had some key projects and had a better-known boss, Rob, often referred to as the father of the aerospace industry because of his creation of the robot. Rob and the CSI Aerospace Corporation seemed to be, nowadays, popular subjects of news coverages and stories in newspapers and magazines.

The only rest periods for Rob were his coffee breaks. With Jack Rosier, his executive vice-president and closest associate, he shared everything on his mind and they had few secrets. Sometimes, he reminisced with Jack over a cup of coffee about his fond memory of Helga. And often, he asked, "Jack, how did I disappoint Helga? *How?*"

"No," Jack replied, "you did not disappoint her."

"Then why did she leave me?"

Jack had no pat answer. As a matter of fact, he had no answer at all. But he said, "Rob, don't blame yourself. I bet that Helga is just as proud of you as she was when she just met you."

Rob's work schedule was staggering: scheduling, conferences, technical briefings, and decision-making as well as some technical assignments on knotty problems which he assumed himself. Even at night when he had already retired to his apartment near the company, he still mulled over some of these knotty problems. But he was very successful with his operations, and all his major projects were on schedule.

The big day finally came: *the launching of Apollo 11 to carry the first man to the moon from Cape Kennedy for a 244,930-mile journey. It was scheduled for July 16. The astronauts for this epic journey were to be Neil A. Armstrong, Edwin A. Aldrin and Michael Collins.*

Rob's chief engineer, Lee Childers, had to supervise his crew and stand by for emergency measures at the Cape during the launching. But Rob and Jack were free agents of their own so far as the launching was concerned, and were invited as VIP guests to witness the historical event.

They stayed in a Cocoa Beach motel for the night of July 15. At 5 a.m. the next morning, they received their coded nametags and were then picked up by a limousine for the Cape Kennedy viewing stands. Their coded nametags entitled them to the *VIPest* VIP viewing stand, where former President Lyndon B. Johnson sat.

When they passed by the former President, he stood up. While President, he had visited Rob's CIS Aerospace Corporation and had a long interesting visit with Rob.

"Hi, Rob," Mr. Johnson said, "how are you? it's nice to see you today."

"How do you do, Mr. President?" Rob responded without quite knowing how to address him. He had never heard any one being addressed as Mr. Former President.

"Rob, you have done a great job. Let me wish you good luck for this venture of historical significance."

"Thank you, sir."

Rob and Jack were seated two rows behind Mr. Johnson.

Finally the countdown and then the takeoff. It was a blastoff in a ball of fire and then a streak of white smoke. Rob followed the rocket in motion for a minute or so. Then it disappeared into the sky yonder.

To Rob's relief it was a successful launching.

For the following two days, Rob followed the activities of the moon journey on television.

On July 20, Neil Armstrong and Edwin Aldrin landed lunar module *Eagle* in the Sea of Tranquility on the moon. And at 10:56 p.m., EDT, Neil A. Armstrong became the first man to set his foot on the moon. As he walked on the lunar surface, he remarked, "One small step for a man, one giant leap for mankind."

Finally, the Apollo command module splashed down in the Pacific Ocean. The astronauts were brought abroad aircraft carrier USS *Hornet* and were greeted by the President.

Robot's reputation grew steadily and rapidly not only in technical and industrial circles. He appeared on the cover of the *Life* magazine with the caption "Mr. Robot, Sr." implying that he had "sired" generations of robots that were sustaining our technological society. Also, there was a long article about him in the magazine, and even his technical keynote speech in Moscow was reproduced word by word and figure by figure.

To his surprise, Rob also received an invitation from the Vice-President of the United States and Speaker of the House to address the joint houses of the Congress, which would also be televised. With a mixed feeling of excitement, pride and humility, he accepted this invitation.

"Mr. Vice-President, Mr. Speaker, distinguished members of both houses of the Congress, ladies and gentlemen," Rob began his speech, "I feel indeed greatly honored and humbled to accept the invitation from Mr. Vice-President and Mr. Speaker to address this body on the subject of *The Innovating Society: Its Technology and Social Implications.*" And he continued:

"There is no doubt that the American society is a changing society. Not long ago, the prevailing belief of our society was that sex was dirty and the environment was clean. For better or worse, these concepts have been changed.

"The American society is not only a changing society. It is also an innovating society, a society that uses its technological prowess to innovate for the physical well-being of its people and for social progress. And the scope of its technological innovations outstrips any of its contemporary societies including all the industrial nations of the world. Today, we shall talk about this innovating society, its technology and social implications."

With this introduction, Rob proceeded to give a historical background. According to him, there are three periods in American history in which technology has taken on different perspectives and forms in our society.

"The first period covers from the Revolutionary War to beginning of the steamboat era about 1807, when so many scientists were also politicians, including Benjamin Franklin (lightning and electricity), George Washington (scientific farming), and Thomas Jefferson (metric system, creation of the patent office). During the second period covering from 1807 to 1942, our emphasis was on applied technology, and for basic research as the foundation for technology we looked to Europe for leadership and answers; and during this period, we had an array of great inventors, pioneers and industrial giants like Thomas Edison, Alexander Bell, Henry Ford, Harvey Firestone, John D. Rockefeller, and the Wright brothers, and we all saw the rise of great industrial centers dotted throughout the landscape. For the third period beginning with the World War II in the early 1940s, we saw unparalleled developments in both basic research and applied technology, and out of the war, the United States became fully engaged in basic research on its own."

Amplifying his statements further he added, "For having won only a sparse dozen or so Nobel Prizes in the forty years up to 1940, United States scientists went on to win some forty in the ensuing years and take first place among the nations."

Having reviewed the history of technology, Rob went on to examine the "anatomy" or "structure" of technology, with its four components: *energy*, *material*, *information*, and *system*, and also to discuss their historical origins.

Now, Rob began to cite some highlights in technological marvels: Space explorations as exemplified by *Mariners 8* and *9* and other programmed space missions, spacecraft and satellites serving communications and navigation, the computer and information systems, DNA (deoxyribonucleic acid) as the prelude to a biological revolution, the integrated circuits (IC) and Large-Scale-Integrated (LSI) systems. When he mentioned the robot and its extensive contributions, he thanked the audience for giving him the credit as the creator of the robot, and then added with emphasis that this credit should really go to a multitude of engineers, scientists and applied mathematicians living in different countries around the world, who had collectively contributed to this accumulation of knowledge in a very vital field for the well-being of mankind. Then he came around to a recent event.

"The moon landing on July 20 was certainly a most appropriate crowning achievement of our time. And I fully agree with Neil A. Armstrong's remark during his moon walk: 'One small step for a man, one giant leap for mankind.' However, in view of the fact that there were thousands of robots in the various electronic and control systems of *Apollo 11* and the mission was a complete success, I wish to paraphase Neil Armstrong's remark to pay our tribute to the robot: *One small step for the robot, a giant leap for mankind.*"

Apparently, Rob wasn't alone in his tribute and praise to the robot. The entire audience stood up and gave a standing ovation to the robot.

Having examined the long-term history of technology, Rob now discussed the trends of technology since the beginning of the century.

"From 1900 to the 1930s, technology was built upon *professional empiricism*, and technology itself was more of an art than a science. From the 1940s to the 1950s, *applied science* became a solid foundation for technology. From the 1960s to the 1970s, technology is becoming more *socially oriented* in the sense that it now has major thrusts in societal problems involving the ecology, environment, energy, information management, transportation, urban problems, etc. In a trend already begun, the 1980s will see technology becoming more *humanistically oriented*, contributing significantly to the mind and body of the human being through biomedical engineering and health-care delivery systems by monitoring and studying human functions and behavior and by incorporating engineering and technological advances into medical practice."

After a brief pause, Rob added, "It is now easy to see that technology is very much *socially involved* with a full range of societal and humanistic problems. Technology is therefore no longer the private domain of the engineers and scientists; and rather, it is the responsibility for all of us, citizens and leaders of our society."

Rob then continued, "You as the leaders of our government and society have the special obligations to formulate far-reaching policies and to appropriate adequate funds in supporting, stimulating and governing the future growth of our technology."

"You *must* give us the guidance through these policies," Rob elaborated, "so that we may exert a concerted technological effort to solve many societal problems, for example, in carrying out the processes of recycling the waste, stopping pollution, and restoring the air and water to accepted standards. And you *must* have the foresight to sustain our technological momentum which is the greatest asset this nation has; for example, the space effort which has just reached a momentary peak of moon landing should be continued. The space effort is an instrument for extending and realizing the imagination and dreams of man. When Christopher Columbus navigated the high seas to find a short sea route to the Indies for gold and spices, he had a dream but did not know that he was to discover this continent. Like Columbus, we have only limited immediate goals for our space explorations. And like Columbus our dreams may lead to a fantastic future unforeseeable at the present by the mortal eyes."

Rob then concluded his speech, "Today, I have presented to you the historical background, anatomy, recent highlights, and treands of technology as well as its social involvements and implications. And I have also challenged you to look after your special obligations and farsighted goals with legislations and appropriations in the name of technology for the long-range well-being of our citizens, our society, our nation and mankind. Our Congress is a great deliberative body, and you are great leaders of our nation. And I have my full confidence in your leadership and wisdom. Allow me now to paraphrase the commercial of one of our great merchandising houses: *We are in good hands!*"

And Rob was greeted with a rousing hand.

Ten

After leaving Rob, Helga also quit her *New York Tribune* job. She moved into a studio apartment in Greenwich Village, upstairs from a store but with a separate stairway and entrance. The studio-bedroom was a large room with a skylight and many large mirrors along the walls. The apartment had had many artistic occupants in the past: painters, dancers, and writers.

Helga had been fighting her conflicting ideas and implosions in her mind, seeking peace and a resolute course to follow.

She'd been in the apartment for a week. Every morning she walked for a block to the corner café for her breakfast. Today, she was sitting at her usual table, and began to order.

"*Helga*, is it you?"

Helga was a little startled as she was not expecting it. A tall long-haired young man was standing by her table, and looking at her, face to face.

"*Rap*!" She stood up, opened her arms and hugged him. "*Rap*, so nice to see you!"

Helga and Rap were actors together in the high school and community theatres in Northern New Jersey, when they were high school classmates. Rap was always the lead, and Helga played the various parts.

"Do you live around here?" Rap asked.

"Yes, just one block this way," she pointed to the north. "I moved in about a week ago."

"I live just the east of here by two blocks."

"What do you do here?"

He grinned, "I am the *star*, the lead man in the show called

The Greatest Show in the Village in the Lisbon Theatre on Lafayette!"

"Great!" And she chuckled, "What an honor to know the star of *The Greatest Show*!"

"The show is a hodgepodge of skits, short acts, ballet and jazz dancing acts." Sensing Helga's impending question he hastened to add, "Nowadays, many popular shows are that way. There is even an enormously successful one on Broadway."

Helga nodded, more acknowledging than approving.

And he continued, "There is also a nude show at the end of the program—it's a dancing act."

"Oh!"

"There is nothing wrong. Like the rest of the show, it's a satire. It is a show of rebellion to the tradition, hypocrisy and injustice in our society. Is marriage such a holy institution that your body is the private property of a man for him to look at and use?"

"That's interesting," Helga responded.

And on her mind, Jane Green's words were now more vivid than ever: "*Even marriage is a form of slavery and fraud. It is a combination of private burlesque, legalized rape, slave labor, forced submission, fraudulent contract, and curtailment of a woman's freedom without any assurance of love from a man. By legal means, we are put in a maximum security prison.*"

"Helga, what do you do now?"

"I have yet to figure out."

Helga and Rap saw each other at breakfast almost every morning. During their discussion, he seemed to be more radical than ever and his view was extremely violent. Helga knew where Rap was each evening—he was in the show and then paraded in the buff in the finale. But she never could find Rap daytime. And he also sometimes disappeared for a day or two, not showing up for breakfast or the show, where his understudy would take over. And he was very mysterious about his whereabouts.

This morning Rap arrived twenty minutes earlier than usual in the Corner Café and waited for Helga. He was somewhat restless. Now, Helga also showed up.

"Helga, will you help me out? Will you be my leading lady in the show? Jo Ann is quitting to get married?"

Conditioned and steeled by Jane Green and Rap Huri with their torrents of words against the tradition, Helga had no problem of making a quick decision. "O.K., I'll help you out. When?"

"You'll be in the show next week. We'll start to rehearse with you tomorrow."

Helga had taken dancing lessons for twelve years in her grade school and high school days. And for many years, she danced different roles in the *Nutcracker Suite* performances during Christmas holidays as well as in jazz numbers, for her local ballet company in Northern New Jersey. But she had not danced lately. To be exact, *lately* meant the last twelve long years.

That evening after having promised Rap to "help out" in the show, she took a shower. Then she walked into her studio-bedroom, still wet with dew-like water drops over her shoulders and torso. She looked into a wall mirror in front of her: her long black shining hair, her sparkling eyes, her exquisite nose. And she put on a smile—a bewitching smile. She was ravishingly beautiful.

When she looked at herself—her body, she was enchanted. In her Missouri girls' dormitory during her college days, *Playboy* magazines were not exactly items of rarity, and she knew a good figure when she saw one. And she imitated the various poses in these magazines: the bunny look, the feline look and the clownish postures. Although she was now thirty, she still looked her beautiful self some ten or twelve years ago. Her firm round breasts with upturned tits—she cupped her palms and lifted them gently. And she smiled at her own image in the mirror.

For a moment, Helga was suddenly seized with guilt for her narcissistic introspection. But she came out real quick. *It was all a conspiracy by men to reject the body. The body is beautiful. We should be proud of it. And I am proud of mine.*

Helga did a cartwheel, a handspring and a handwalk, and looked into the mirrors on the wall for the images and images' images (she was also a tumbler in her early dancing training). The multi-imaged movements in the mirrors fascinated her.

Now, she felt ready to practice her dancing steps just to warm up for next week. Pirouettes, arabesques, grands and tour jetés as well as various variations. It was beautiful to watch the kaleidoscopic displays on the wall mirrors.

Helga's first night on the stage finally came. She was not in any of the skits or short acts. But she was in most of the dancing numbers. And she did them beautifully.

The theatre was almost fully packed. It was small and the audience in the first row could almost touch your body with their hands. Between her numbers, she thought about her nude act and was nervous and stage-frightened. Even since she was ten, she had never paraded herself *au naturel*, even before her father or Aunt Anita. And now——.

The final dancing numbers now came. Just in a matter of minutes her self-consciousness melted away. She felt that she was the Joan of Arc for the women's cause, tearing away the tradition and charging forward for the independence of women in the male-dominated society. And she danced gracefully with a triumphant smile constantly on her face.

In her younger days, she saw Rodgers and Hammerstein's *Oklahoma*! both on Broadway and the movie, with her father, and she loved it. She thought that the song, "*Oh What a Beautiful Mornin'*!" was the ultimate glory of any cause. Now, she was triumphing for the cause of women. And now in a mental picture she saw herself as Laurey, and Rap as Curley on a horse, in an Oklahoman cornfield singing:

> "*Oh, what a beautiful mornin'!*
> *Oh, what a beautiful day!*
> *I got a beautiful feelin'*
> *Ev'rythin's goin' my way.*"

Naturally, they were singing in the buff.

The show went on month after month to near-sold-out capacity, and Helga kept on searching the meaning of her fulfillment.

Eleven

Helga had noticed that whenever Rap was absent from the show and breakfast for a day or two there was the coincidence of some news about the bombing of a recruiting station of an armed service, a bank, or a military installation in the East or Midwest, with a "straw cross" left at the bombed site. The police in New York City did arrest a suspect after one of these bombings, and learned that the organization responsible for these violent acts was known as the VAR, the Violent Actions for Revolution, and was active in the Eastern and Midwestern states.

Rap was absent from the show and breakfast for four days. Helga was having breakfast at the Corner Café alone, reading the *New York Times*. The headline of the paper read:

MID-WESTERN UNIVERSITY BOMBED
3 RESEARCHERS DIED

In reading the details, Helga learned that a truck with dynamite timeset to explode at midnight was parked outside Mallory Hall of the campus, in which mathematical research sponsored by the Army was conducted. The dynamite went off at 11:58 P.M., blasted off one side of the huge building, and killed a research professor and two graduate students instantly. A straw cross was found in front of the dynamited building. Helga was indignant.

Before Helga finished her breakfast, Rap showed up.

"Did you do it, Rap?" Helga angrily asked, pointing her finger at the headline of the newspaper.

"*Yes*," Rap answered with some air of pride.

"Why did you do it?"

"Our society is corrupt, and we have to rebuild from scratch."

"Was it dirty of you and your Mafia to have killed three innocent people under the pretense of reform?"

Rap was enraged. "Helga, they are only pigs. Why should you care?"

"Rap, they are a professor and two students doing mathematical research!"

"Helga, don't be naive! They are the instruments of American imperialism. Besides, what could the lives of a couple of people mean to you, to me or anyone?"

"Rap, they mean a lot to me. No reform is worthwhile if a single life is to be sacrificed. I want a *just* society in which reasons prevail, and not one in which wrongs are to be rectified with wrongs. Two wrongs do not make it right."

"Damn it, Helga, you are a softie!" And Rap stormed out of the café.

Rap quit the show at the Lisbon Theatre. And he also stopped coming to the Corner Café. Helga prayed that Rap would change. *God bless him*!

Two months later, Helga read in the paper this headline:

TOWN HOUSE DEMOLISHED BY BLAST
SEVEN YOUNG RADICALS KILLED

And the fine print read as follows: "In New York, a blast demolished a fashionable Greenwich Village townhouse, and police unearthed evidence that it had been caused by the accidental detonation of a câché of bombs that a band of young radicals has been putting together in the basement. Seven bodies were uncovered and only three men were immediately identified: Leo Adams, Rap Huri, and Jack Narud."

After Rap's funeral, Helga quit her dancing job at the Lisbon Theatre.

For years, Helga and her father Frank McGee always tried to have a lunch together once a week. This week, they instead arranged a dinner at the Archor Restaurant in the Village, where waitresses in red shipboard short uniforms served the drinks and food.

Frank had also kept in regular, if infrequent, touch with Rob. He was Helga's only source of information about Rob.

"Dad, how is Rob?"

"He is fine. I had dinner with him last Thursday. He is working as hard as ever."

"I saw him on television when he addressed the joint houses of the Congress. He looked fine to me."

Frank was happy to note that her interest in Rob was still alive if somewhat dormant.

Helga had told Frank about many of Rap's earlier activities. Now after telling him about Rap's latest episodes, she asked: "Dad, why did he have to kill people to reform the society?"

Having read the *Life* magazine article about Rob and the reprint of his keynote speech about the robot in Moscow, he said: "Helga, do you remember in Rob's speech in Moscow that if the robot receives a certain 'command,' say, a step function" (*See Fig. 8b in Chapter 6*), "nature does not allow the effect or response to be entirely identical to the command, and instead the 'effect' only approximates the command and has overshoots?" (*See Fig. 8c*). "The overshoots are the excesses like the killings. As long as the robot, namely, the society in this case, is *stable*, the overshoots will eventually *die out*. And this is merely a physical law, dictated by nature or the physical constraint of the robot, for us to obey."

"Dad, I didn't know that you are also one of Rob's disciples."

"Helga, Rob and I also had many discussions. He told me that by his concept robots themselves can be and often are components of another robot of a higher order (which has its own feedback loop and system model), and these robots of higher order can be and often are components of another robot of an even higher order. The sophisticated electronic and control systems now in existence often contain robots of very high orders."

Helga listened with great interest. And Frank continued: "Think about the implications of this concept. We, human beings, can be the basic robots. With good 'citizen' robots like us as components, our governments and societies of different levels can be robots of higher orders of different levels."

"We design robots to be stable and provide good performance" (*See Rob's keynote speech in Chapter 6*), "and a good robot, so designed, will have all its robotosyncratic virtues. If we set out to design good 'citizen' robots, and then good 'government' or 'society' robots of higher orders, we are more likely to be able to design or produce a good 'world' robot. Who doesn't like to live in a world (call it robot or not) that reacts to you in a broad-minded, even-tempered, agile, stable, visionary, resolute and reliable manner?"

You are a good philosopher," she grinned, "but, Dad, tonight I really want to say *good-bye* to you. I am leaving for California."

And there was silence in the ensuing moment.

"Helga, I am puzzled. Have you gone through enough soul-searching to realize where your future belongs? Isn't it time that you return to Rob? You need each other!"

"Dad, I am still fighting the conflicts within me. I have conflicting urges, the urge to dedicate myself to social causes and women's equality, and the urge to play the role of a good wife. And I cannot face Rob without first resolving this basic conflict."

"This is fine. Let's now return to the robot. The robot too has conflicting urges in terms positive errors (*e.g.*, E_a, E_c . . .*in Fig 8c*) pulling the effect (or response) y up and negative errors (*e.g.*, E_b, E_d, . . .*in Fig. 8c*) pulling the effect y down. However a wise (stable) robot will use these conflicting urges to damp out the overshoots and cause the effect y to follow the command x *(see Fig. 8c)*. In other ways, these conflicting urges will keep the effect or response on target."

"Dad, it is beautiful."

Frank then pleaded, "Helga, when the overshoots in your system dampen out, will you promise to return to Rob?"

Enveloping Frank's hand with hers on the table, she said, "I promise. But, Dad, I need first to be all by myself and think things over."

He nodded his head.

Planting a kiss on his cheek, she forced a smile on her face and said, "Dad, you may not hear from me for a long, long time."

Twelve

Following the moon landing, a recession set in and the space budget for the nation was drastically scaled down. The CSI Aerospace Corporation lost within a year many of its major space contracts. It laid off one third of its original force of 9,000. The major portion of its remaining work force now worked on the various projects for the SST (Supersonic Transport), which seemed to be secure at the moment.

However, in the ensuing six months, the SST became a great environmental issue, and SST battles were fought and lost one after another in the Congress. With the latest cutoff of federal funds for the SST program, it seemed certain that 5,000 of the 6,000 work force in the CSI would have to be let go.

On a gray Friday afternoon, the CSI board of directors met and decided to absorb the CSI Aerospace Corporation into the parent organization, the Communication and Systems, Inc. They offered Rob a vice-presidency in the parent organization.

It was a hard decision for Rob to make. He really couldn't face the fact that 8,000 of his 9,000 colleagues had been or would be laid off and that he would continue to enjoy the amenities of the executive suite. In two days he made up his decision. He too was going to leave the company, and resigned, effective in thirty days in order to take care of the scheduled closedown and merger with the parent company.

On Rob's first day of unemployment, he went back to his Hideaway. After sunset, he sat on the beach facing the five rocks.

There was a full moon just like the night he and Helga had

their first rendezvous here some seven years ago.

He looked at the five rocks. They were the symbols of his childhood, and the good omens for his creation of the robot and for his encounter and discovery of Helga.

For the creation of the robot, it was like a dream, and in waking up he felt like a poorly designed robot in having lost his stability. Perhaps he was not properly designed.

For Helga, it was a joyous experience but——. As he looked into the water, he saw the reflection of Helga and himself standing on one of the five rocks:

They embraced and kissed. And she enveloped his hands with a terrific squeeze and said, "Love is not a comma, but a period. A helluva grip of a period."

Rob smiled a smile in reminiscence. And other mental images began to appear: *their nude swimming and frolicking in the Long Island Sound, their marriage by the sea, the declaration of Helga's Period on Mount Avila, their make-believe moon, the romantic moments at the Red Square, the Kremlin tour, the Novodevichy Convent and their tender moments.*

"Was this the end of the line?" Rob asked of himself.

A hand seemed to rest on Rob's shoulder, and he turned around.

"*Rob! Rob!*" She threw her arms around him.

They hugged tightly and kissed.

"*Helga! Helga! Helga!*" He kept on repeating.

They looked at each other, and touched each other's face by hand with gentle strokes just to make sure that it was real.

Helga now held Rob's hands and gave them a terrific squeeze. And both of them chimed in, "Love is not a comma, but a period. A helluva grip of a period!"

Appendix A

Study of Network Responses with the Fourier Transform

9.1. Response of an idealized low-pass four-terminal network subject to a unit-step excitation

A. Description of Problem

A four-terminal network, say an amplifier or a passive network, is usually rated according to its steady-state and transient performances. To determine the transient performance of a network, a unit-step function as depicted in Fig. 9.1a,

$$u(t) = \begin{cases} 1 & t \geq 0 \\ 0 & t < 0 \end{cases} \tag{9.1}$$

or a step function, which is a unit-step function multiplied by a constant, is often used as the excitation $e_1(t)$ to the network, as indicated in Fig. 9.1b. The response $e_2(t)$ due to the step-function excitation is usually studied with particular reference to:

1. *The rise transient response and its rise time.* The definition of *rise time* varies among authors. It may be defined as the time required for the transient to reach nine-tenths of its maximum value, e.g., the time interval OF in Fig. 9.1c.

2. *The persistence of the response or "staying power."* The persistence of the response may be measured by the decay time, which in turn can be defined as the time required for the transient to decay from the maximum to one-tenth of this value. For example, the time interval CG in Fig. 9.1c is the decay time. The larger the decay time, the better the staying power of the response.

The study of transient performance is particularly important for video amplifiers and pulse and switching circuits.

We shall find in Art. 9.4 that the steady-state and transient performances of a network are closely related. Knowing the transient response of a network subject to a step-function excitation, we can predict the steady-state response subject to a sinusoidal excitation.

Problem under investigation. We shall investigate the response $e_2(t)$ of a four-terminal low-pass network having idealized transmission characteristics as indicated in Fig. 8.5a and b subject to a unit-step excitation $e_1(t) = u(t)$. Let us assume $K = 1$ in Fig. 8.5a. This means that the network will pass freely frequencies $\omega \leq \omega_c$ and cut out all $\omega > \omega_c$. The d-c component in the excitation is identified with $\omega = 0$ and therefore will appear with its original magnitude in the response.

The response $e_2(t)$ in Fig. 9.1c decayed because it lacked a d-c component. The network transmission characteristic $G_0(\omega)$ must have the relation $G_0(0) = 0$ and must eliminate the d-c component in the response.

For our present problem, we have a transmission characteristic $G_0(\omega) = 1$ for $-\omega_c \leq \omega \leq \omega_c$. This means that the d-c component in the excitation $e_1(t)$ will

W.H. Chen, "The Analysis of Linear Systems," McGraw-Hill Book Co., New York, 1963, pp. 213-220, plus some reference information (pp. 200, 209-210)

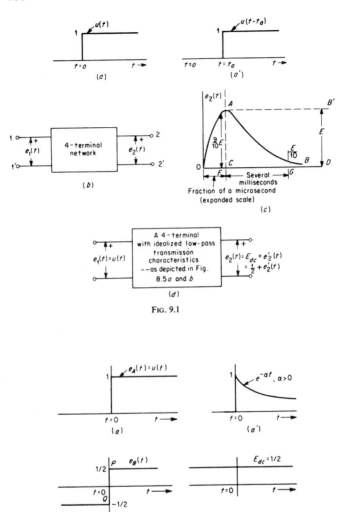

FIG. 9.1

FIG. 9.2

appear in the response $e_2(t)$ with the original magnitude. Therefore, the response $e_2(t)$ will "stay up" and will follow the general shape of OAB' in Fig. 9.1c. We have no decay transient for the test of "staying power," but only a *rise transient*, to study. We shall investigate the rise transient for its (1) shape, i.e., waveform, (2) rise time, and (3) time delay as introduced by the four-terminal network.

A remark about the d-c component of a signal $e(t)$ **and the Fourier transform** $g(\omega)$ **of this signal.** From earlier discussions, we know that (1) we may find the Fourier transform or spectrum function $g(\omega)$ of a given signal function $e(t)$ with the aid of (8.3b) and (2) we may derive the signal function $e(t)$ from the Fourier transform $g(\omega)$ with the aid of (8.3a).

Let us assume $e_A(t) = u(t)$ as in Fig. 9.2a and find its Fourier transform $g(\omega)$ with the aid of (8.3b). If we use this $g(\omega)$ in (8.3a), we shall find with appropriate limiting processes, instead of $e_A(t)$, a new signal function $e_B(t)$ having an approximate representation as depicted in Fig. 9.2b. It is obvious that

$$e_A(t) = e_B(t) + E_{dc} = e_B(t) + \tfrac{1}{2} \qquad (9.2)$$

where $E_{dc} = \tfrac{1}{2}$ (Fig. 9.2c) is the d-c component in $e_A(t)$. The Fourier transform $g(\omega)$ apparently does *not* include the d-c component in the signal function $e(t)$. In a network problem solved by Fourier transforms, we must treat the d-c component separately.

A remark about our present problem. We shall investigate the response $e_2(t)$ of a four-terminal network having idealized characteristics as depicted in Fig. 8.5a and b subject to a unit-step excitation $e_1(t) = u(t)$. According to the above remark, we must treat the d-c component separately. The d-c component in the excitation $e_1(t) = u(t)$ is obviously $\tfrac{1}{2}$ (say in volts) at any instant (including the negative time axis) as depicted in Fig. 9.2c; the d-c component in the response $e_2(t)$ for a transmission characteristic $G_0(\omega) = 1$ for $-\omega_c \leq \omega \leq \omega_c$ must be $E_{dc} = 1 \times \tfrac{1}{2} = \tfrac{1}{2}$. The response must then be

$$e_2(t) = E_{dc} + e_2'(t) = \tfrac{1}{2} + e_2'(t) \qquad (9.3)$$

where $e_2'(t)$ is the a-c response subject to the excitation $e_1(t)$ as indicated in Fig. 9.1. $e_2'(t)$ can be obtained with the procedure in (8.34).

B. Procedure for Finding the Response Subject to a Unit-step Excitation

We shall find the response for the problem described above and in Fig. 9.1d using the procedure in (8.34).

Fourier transform of a unit-step excitation. Strictly speaking, the Fourier transform of a unit-step excitation does not exist [this will be discussed in an illustration associated with Eq. (11.16a)]. We shall use a limiting process to get around this difficulty for the time being and introduce the Laplace transform in Art. 11.3 as a generalization of the Fourier transform to take care of many functions whose Fourier transforms do not exist.

The assumption that a unit-step excitation is an exponential excitation $e^{-\alpha t}$ with extremely small $\alpha > 0$, that is, a limiting case, should be good enough for *practical* circuit applications.

Let us now consider a unit-step function as the limit of an exponential function as depicted in Fig. 9.2a', i.e.,

$$e_1(t) = u(t) = \begin{cases} \lim_{\alpha \to 0} e^{-\alpha t} & \alpha > 0, t \geq 0 \\ 0 & t < 0 \end{cases} \qquad (9.4a)$$

and then find its Fourier transform with the aid of (8.3b) and a limiting process:

$$\begin{aligned} g_1(\omega) &= \frac{1}{2\pi} \int_{-\infty}^{\infty} e_1(t) e^{-j\omega t}\, dt = \frac{1}{2\pi}\left(\int_{-\infty}^{0} + \int_{0}^{\infty} \right) \\ &= 0 + \lim_{\alpha \to 0} \frac{1}{2\pi} \int_{0}^{\infty} e^{-\alpha t} e^{-j\omega t}\, dt \\ &= \lim_{\alpha \to 0} \frac{1}{2\pi} \frac{1}{\alpha + j\omega} = \frac{1}{2\pi} \frac{1}{j\omega} \end{aligned} \qquad (9.4b)$$

Fourier transform of the response. Using (8.35a) and the prescribed transmission characteristics for $K = 1$ in Fig. 8.5a and b, we have

$$g_2(\omega) = \begin{cases} \dfrac{1}{2\pi} \dfrac{1}{j\omega}(1 e^{-j\omega t_d}) & -\omega_c \leq \omega \leq \omega_c \\ 0 & \omega < -\omega_c,\ \omega_c < \omega \end{cases} \qquad (9.5a)$$

Network response. The response has the form [Eq. (9.3)]

$$e_2(t) = \tfrac{1}{2} + e_2'(t) \qquad (9.5b)$$

and $e_2'(t)$ may be obtained with the aid of (8.35b).

$$\begin{aligned} e_2(t) &= \frac{1}{2} + \int_{-\omega_c}^{\omega_c} \frac{1}{2\pi} \frac{1}{j\omega} e^{j\omega(t - t_d)}\, d\omega \\ &= \frac{1}{2} + \frac{1}{2\pi} \int_{-\omega_c}^{\omega_c} \frac{1}{j\omega} [\cos \omega(t - t_d) + j \sin \omega(t - t_d)]\, d\omega \end{aligned} \qquad (9.6a)$$

$$e_2(t) = \frac{1}{2} + 0 + \frac{1}{2\pi} \int_{-\omega_c}^{\omega_c} \frac{\sin \omega(t - t_d)}{\omega} \, d\omega \dagger$$

$$= \frac{1}{2} + 0 + \frac{2}{2\pi} \int_0^{\omega_c} \frac{\sin \omega(t - t_d)}{\omega} \, d\omega \ddagger \tag{9.6b}$$

Letting
$$U = \omega(t - t_d)$$
$$x = U \big|_{\omega = \omega_c} = \omega_c(t - t_d) \tag{9.7a}$$

and
$$\frac{d\omega}{\omega} = \frac{dU/(t - t_d)}{U/(t - t_d)} = \frac{dU}{U} \tag{9.7b}$$

we may rewrite (9.6b) as

$$e_2(t) = \frac{1}{2} + \frac{1}{\pi} \int_0^x \frac{\sin U}{U} \, dU = \frac{1}{2} + \frac{1}{\pi} \, \text{Si} \, x \tag{9.6c}$$

$$e_2(t) = \frac{1}{2} + \frac{1}{\pi} \, \text{Si} \, [\omega_c(t - t_d)] \tag{9.6d}$$

† $G(\omega) = \cos \omega(t - t_d)/j\omega$ is an *odd* function of ω according to definition (8.10b), and $\int_{-\omega_c}^{\omega_c} G(\omega) \, d\omega = 0$. For reference, see $\int_{-t_a}^{t_a} f(t) \, dt = 0$ as the area under the curve in Fig. 7.7a between the limits $-t_a$ and t_a.

‡ $F(\omega) = \sin \omega(t - t_d)/\omega$ is an even function of ω according to definition (8.10a), and $\int_{-\omega_c}^{\omega_c} F(\omega) \, d\omega = 2 \int_0^{\omega_c} F(\omega) \, d\omega$. For reference, see $\int_{-\omega_c}^{\omega_c} g(\omega) \, d\omega = 2 \int_0^{\omega_c} g(\omega) \, d\omega$ as the area under the curve in Fig. 8.4a between the limits $-\omega_c$ and ω_c for any ω_c.

where Si x is the *sine integral* of x, defined as the integral of a sampling function Sa $U = (\sin U)/U$:

$$\text{Si} \, x = \int_0^x \frac{\sin U}{U} \, dU \tag{9.7c}$$

To plot the response $e_2(t)$, it is only necessary to use Eq. (9.6d) and a table of sine integrals.† A typical response is given in Fig. 9.3d.

Response waveform. To study the response $e_2(t)$ as represented in (9.6d), we may (1) plot the sampling function Sa U in Fig. 9.3a, (2) plot the sine integral Si x in Fig. 9.3b with Si x_1 equal to the area under the Sa U curve in Fig. 9.3a from $U = 0$ to $U = x_1$ for any value of x_1, Si x being defined in (9.7c), and (3) change the scales along both axes to obtain the horizontal axis and add $\frac{1}{2}$ to the vertical representation, obtaining the response $e_2(t)$ in Fig. 9.3d.

Comparing the response $e_2(t)$ in Fig. 9.3d with the unit-step excitation in Fig. 9.1a, we note that:

1. A time delay t_d in seconds is introduced in the response.
2. A time interval is required for the buildup of the transient response.
3. Overshoots or oscillations are introduced in the response.

These three phenomena are introduced by the four-terminal network having idealized transmission characteristics as depicted in Fig. 8.5a and b.

C. Buildup Time of the Response

The buildup time of the transient response $e_2(t)$ as indicated in Fig. 9.3d is the time required for the transient to proceed from F to G, that is, from its zero value to its steady-state value.

Let us reproduce the transient response $e_2(t)$ in Fig. 9.4, draw a straight line $F'AG'$ tangent to A, and construct a right triangle $F'H'G'$. The time interval t_B, the "length" of side $F'H'$ of the triangle, is an approximation of the buildup. We now wish to find t_B.

The slope $1/t_B$ is also the derivative $d[e_2(t)]/dt$ at $t = t_d$ (i.e., at A in Fig. 9.4). With reference to (9.7), we have

$$\frac{1}{t_B} = \frac{d}{dt} \, [e_2(t)]_{t=t_d} = \frac{d}{dt} \left(\frac{1}{2} + \frac{1}{\pi} \, \text{Si} \, x \right)_{t=t_d} \tag{9.8a}$$

$$\frac{1}{t_B} = \frac{1}{\pi} \frac{d(\text{Si} \, x)}{dx} \bigg|_{t=t_d} \frac{dx}{dt} \bigg|_{t=t_d} = \frac{l_o}{\pi} (1)(\omega_c) \tag{9.8b}$$

† See E. Jahnke and F. Emde, "Tables of Functions," Dover Publications, Inc., New York, 1945, for such a table.

Equation (9.8b) results from

$$\frac{d}{dx}\left(\text{Si } x\right)_{\substack{t=t_d \\ x=\omega_c(t-t_d)}} = \frac{d}{dx}\left(\int_0^x \frac{\sin U}{U} \, dU\right)_{\substack{t=t_d \\ U=\omega(t-t_d),\, \omega=\omega_c}}$$

$$= \frac{\sin U}{U}\bigg|_{U=0} = 1 \qquad (9.9a)$$

$$\frac{dx}{dt} = \frac{d}{dt}\left[\omega_o(t-t_d)\right] = \omega_c \qquad (9.9b)$$

From (9.8), we find the *buildup time*,

$$t_B = \frac{\pi}{\omega_c} = \frac{\pi}{2\pi f_c} = \frac{1}{2f_c} \qquad (9.10)$$

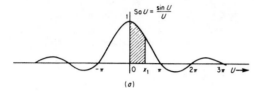

(a)

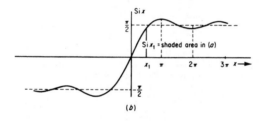

(b)

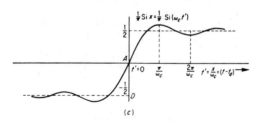

(c)

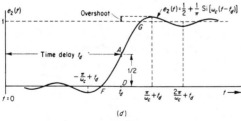

(d)

FIG. 9.3

where ω_c is the cutoff frequency in radians per second, and f_c in cycles per second, of the idealized low-pass transmission characteristic as depicted in Fig. 8.5a.

As an illustration, a buildup time of 1 msec is expected for a cutoff frequency of 500 cps from (9.10). Since (9.10) was derived with a low-pass network with idealized transmission characteristics, it is only an approximate equation for application to actual physical networks.

D. Remarks about the Response of a Network Subject to a Given Excitation

We have demonstrated above that, for a unit-step-function excitation $e_1(t) = u(t)$ to a four-terminal low-pass network with idealized transmission characteristics, the response $e_2(t)$ in Fig. 9.4 is usually regarded as a transient response. The unit-step excitation is not a repeating function as defined in (7.1). We may in general consider the network response subject to a nonrepeating function as a *transient response*.

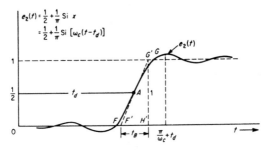

FIG. 9.4

Finding the response of a two-terminal network subject to a given excitation. If the excitation is a repeating function, we may find the steady-state response according to the procedure in Art. 8.2B. If the excitation is a nonrepeating function, we may find the transient response using Laplace-transform methods to be discussed in Chaps. 12 and 13.

Finding the response of a four-terminal network subject to a repeating excitation. For a repeating function as excitation, the steady-state response may be found according to the procedure in Art. 8.2C.

Finding the transient response of a four-terminal network subject to a nonrepeating excitation. There are two approaches to this problem, depending upon how the four-terminal network is characterized.

If the transfer function $G(s)$ of the network is known, we may use the Laplace-transform method, as discussed in Chaps. 12 and 13, to find the transient response.

If the transmission characteristics of the network in the form of the gain characteristic $G_0(\omega)$ and the phase characteristic $\theta(\omega)$ are available, the method of Fourier transforms as described in (8.34) and demonstrated in Art. 9.1B may be used in the determination of the transient response. The transfer function $G(s)$ and the transmission characteristics $G_0(\omega)$ and $\theta(\omega)$ are related in (3.39). Knowing $G(s)$, we can readily find $G_0(\omega)$ and $\theta(\omega)$. But the converse is not true. For a network (1) having a complicated network configuration or a large number of circuit elements, (2) containing distributed circuit elements such as transmission lines or significant wiring capacitances, or (3) whose circuit diagram is unavailable, $G(s)$ is difficult to determine, while $G_0(\omega)$ and $\theta(\omega)$ are easy to measure and can be readily plotted as graphs. With $G_0(\omega)$ and $\theta(\omega)$ available, the transient response may readily be obtained.

If $G_0(\omega)$ and $\theta(\omega)$ are available in graphical form, we may approximate them as functions of ω. For example, we may consider $G_0(\omega)$ as three linear segments, for the positive frequencies, and let $G_0(\omega) = f_1(\omega)$ for $0 \leq \omega \leq \omega_1$, $G_0(\omega) = f_2(\omega)$ for $\omega_1 < \omega \leq \omega_2$, and $G_0(\omega) = f_3(\omega)$ for $\omega_2 < \omega < \infty$, where $f_1(\omega)$, $f_2(\omega)$, and $f_3(\omega)$ are three equations for straight lines [for example, $f_1(\omega) = m_1\omega + b_1$]. Since $G_0(\omega)$ is an even function of ω, we may easily include the negative frequencies. Equation (8.33b) may then be used in the evaluation of the transient response.

Knowing the gain characteristic $G_0(\omega)$, but not knowing the phase characteristic $\theta(\omega)$, we may find the approximate waveform of the transient response. We note in our illustration in Art. 9.1B that a "linear" phase characteristic $\theta(\omega) = \omega t_d$ (Fig.

8.5*b*) introduces a time delay t_d into the transient response $e_2(t)$ to a unit-step excitation (Fig. 9.4). . For any nonrepeating excitation, we may then use a transmission characteristic $G(j\omega) = G_0(\omega)e^{-j\theta(\omega)} = G_0(\omega)$, where $G_0(\omega)$ is the given gain characteristic and $\theta(\omega) \equiv 0$ is assumed, and follow (8.34) to find the transient response without time delay as $\theta(\omega) = \omega t_d \equiv 0$ and $t_d \equiv 0$. The transient response thus obtained has the approximate waveform of the true response, with some error introduced due to the "nonlinearity" of the phase characteristic.

Chapter 8

Reference Information
For Appendix A Text

Definitions

Fourier integral:‡

$$e(t) = \int_{-\infty}^{+\infty} g(\omega)e^{j\omega t}\, d\omega \qquad (8.3a)$$

Equation (8.2*a*) becomes

Fourier transform:§

$$g(\omega) = \frac{1}{2\pi} \int_{-\infty}^{+\infty} e(t)\, e^{-j\omega t}\, dt \qquad (8.3b)$$

Idealized Transmission Characteristics

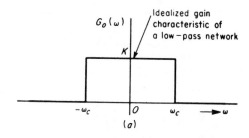

(a)

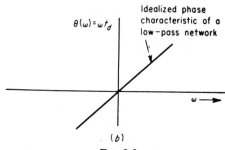

(b)

FIG. 8.5

Chapter 8

Reference Information
For Appendix A Text

Response $e_2(t)$ subject to a nonrepeating function $e_1(t)$ as excitation. We may find the Fourier transform of the excitation $e_1(t)$ in the form of (8.3b),

$$g_1(\omega) = \frac{1}{2\pi} \int_{-\infty}^{+\infty} e_1(t)e^{-j\omega t}\, dt \tag{8.32}$$

which represents the complex amplitudes of the frequency components of the excitation $e_1(t)$ on a relative basis. Since the transmission characteristic $G(j\omega) = G_0(\omega)e^{-j\theta(\omega)}$ relates the relative complex amplitude of each frequency component in the response to that in the excitation, we have

$$g_2(\omega) = g_1(\omega)G(j\omega) = g_1(\omega)G_0(\omega)e^{-j\theta(\omega)} \tag{8.33a}$$

which is the Fourier transform of the response $e_2(t)$. To find the response, we substitute (8.33a) into (8.3a):

$$e_2(t) = \int_{-\infty}^{+\infty} g_2(\omega)e^{j\omega t}\, d\omega = \int_{-\infty}^{+\infty} g_1(\omega)G_0(\omega)e^{-j\theta(\omega)}e^{j\omega t}\, d\omega \tag{8.33b}$$

Summarizing the above procedure, we have

$$\begin{array}{l} \textit{Time domain:} \quad \begin{bmatrix} \text{Given excitation } e_1(t) \\ \text{in Fig. 8.3}a \end{bmatrix} \xrightarrow{\text{Problem}} \begin{bmatrix} \text{Find response } e_2(t) \end{bmatrix} \\[2ex] \qquad\qquad\qquad \downarrow \begin{array}{l}\text{Step 1:}\\ \text{Use (8.32)}\end{array} \qquad\qquad\qquad\qquad \uparrow \begin{array}{l}\text{Step 3:}\\ \text{Use (8.33}b)\end{array} \\[2ex] \textit{Frequency domain:} \begin{bmatrix} \text{Fourier transform of} \\ \text{excitation } g_1(\omega) \end{bmatrix} \xrightarrow[\text{Use (8.33}a)]{\text{Step 2:}} \begin{bmatrix} \text{Fourier transform} \\ \text{of response } g_2(\omega) \end{bmatrix} \end{array} \tag{8.34}$$

For illustration, let us consider a low-pass network with idealized *transmission characteristics* as in Fig. 8.5a and b. For an excitation $e_1(t)$ having a Fourier transform $g_1(\omega)$ as represented in (8.32), the Fourier transform of the response is found to be

$$g_2(\omega) = \begin{cases} g_1(\omega)Ke^{-j\omega t_d} & -\omega_c \leq \omega \leq \omega_c \\ 0 & \omega < -\omega_c,\ \omega_c < \omega \end{cases} \tag{8.35a}$$

The response $e_2(t)$ will then have the form

$$e_2(t) = \int_{-\infty}^{+\infty} g_2(\omega)e^{j\omega t}\, d\omega = \int_{-\omega_c}^{\omega_c} g_1(\omega)Ke^{-j\omega t_d}e^{j\omega t}\, d\omega$$

$$= K\int_{-\omega_c}^{\omega_c} g_1(\omega)e^{j\omega(t-t_d)}\, d\omega \tag{8.35b}$$

Appendix B

20.4. Type 0 closed-loop control systems—regulator systems

We shall give two illustrations of type 0 systems. A type 0 system is often a regulator system, e.g., a "voltage regulator" or "speed regulator," used to regulate a physical quantity, e.g., voltage speed. A type 0 closed-loop system, as compared with an open-loop system, performs better in the sense that it is less susceptible to (1) noise or disturbance, (2) frequency distortion, and (3) unreliability of individual system components. We shall discuss these points in the following illustrations.

A. Illustration 1: Negative-feedback Amplifier

Description of circuit arrangement. The a-c circuit of a two-stage RC-coupled amplifier with feedback loop is depicted in Fig. 20.8a. The 1:1 turns-ratio transformer is purposely introduced so that the schematic of this amplifier (Fig. 20.8b) can be compared with a typical schematic of a closed-loop system (Fig. 20.8c). In practice, this transformer is not used, and the phase reversal of βE_0 can be taken care of by other arrangements.†

Open-loop and overall transfer functions. We note the similarity between the schematics in Fig. 20.8b and c. Corresponding to the open-loop transfer function $G(s)H(s)$ in Fig. 20.3b, we now have

Open-loop transfer function:

$$G(s)H(s) = G(s)\beta \qquad (20.44)$$

Corresponding to the overall transfer function

$$G'(s) = \frac{C(s)}{R(s)} = \frac{G(s)}{1 + G(s)H(s)}$$

we now have

Overall transfer function:

$$G'(s) = \frac{E_o}{E_{\text{in}}} = \frac{G(s)}{1 + G(s)\beta} = \frac{G_1(s)G_2(s)}{1 + G_1(s)G_2(s)\beta} \qquad (20.45)$$

which is also called the *closed-loop gain* of the amplifier.

Voltage relations. $G(s)$ and $G'(s)$, for $s = j\omega$, are functions of the frequency ω. We shall now choose a typical frequency ω_1 and use numerical values $G = G(j\omega_1)$ and $G' = G'(j\omega_1)$ to illustrate the effects of negative feedback. Let us assume

$G_1 = 10\underline{/180°} = -10 = $ first-stage amplifier gain

$G_2 = 10\underline{/180°} = -10 = $ second-stage amplifier gain

$G = G_1G_2 = 100\underline{/360°} = 100 = $ two-stage gain without feedback (20.46a)

$\beta = \dfrac{R_2}{R_1 + R_2} = 0.09$

$E_{\text{in}} = 10\underline{/0°} = 10$ volts $= $ signal input

---

† See J. D. Ryder, "Electronic Engineering Principles," fig. 9.22c, p. 258, Prentice-Hall, Inc., Englewood Cliffs, N.J., 1952.

W.H. Chen, "The Analysis of Linear Systems,"
McGraw-Hill Book Co., New York, 1963, pp. 537-545

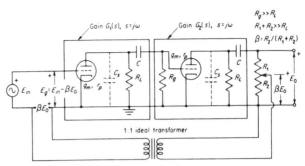

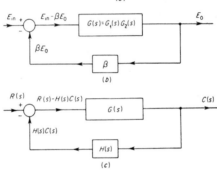

FIG. 20.8

Through computation, we find

$$G' = \frac{100}{1 + 0.09 \times 100} = 10$$

$$E_o = G'E_{in} = 100 \text{ volts}$$

$$-\beta E_o = -9 \text{ volts}$$

(20.46b)

The voltage relations are depicted in Fig. 20.9. We note that the feedback voltage $-\beta E_o$ opposes the signal input E_{in} in polarity, and $|E_o| < |E_{in}|$. We shall say that we now have a *negative-feedback* amplifier. In general, when $|G'| < |G|$, we have negative feedback.

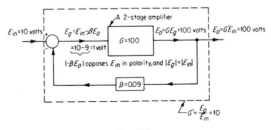

FIG. 20.9

Effects of negative feedback. We shall now try to summarize some of the effects of negative feedback.

1. *Reduction in Gain.* We have demonstrated in Fig. 20.9 that, for $G = 100$ without feedback, we have $G' = 10$ with negative feedback. This means that we have a reduction in gain due to the negative feedback. However, negative feedback also offers many attractive features to offset this disadvantageous gain reduction.

Usually we are not too concerned with the reduction in gain introduced by negative feedback. Additional amplifier stages may be used to make up the reduction.

2. *Closed-loop Gain G' Made Independent of Fluctuations.* For $s = j\omega$ and $|G(j\omega)\beta| \gg 1$, we may approximate Eq. (20.45) with

$$G'(j\omega) = \frac{G(j\omega)}{1 + G(j\omega)\beta} \cong \frac{G(j\omega)}{G(j\omega)\beta} = \frac{1}{\beta} \qquad (20.47)$$

This means that, under the condition $|G(j\omega)\beta| \gg 1$, or a very large forward control-element gain $|G(j\omega)| \gg |1/\beta|$, we can make the closed-loop gain $G'(j\omega)$ at any frequency ω almost *constant* and independent of any fluctuations in the system (e.g., fluctuations in the d-c supply voltage due to unfiltered a-c components, fluctuations in circuit parameters due to aging or unreliability of circuit elements, etc.). This implies that we may use inexpensive nonprecision circuit elements to build a reliable negative-feedback amplifier.

3. *Reduction of Noise in Output as a Negative-feedback Effect.* For the amplifier depicted in Figs. 20.8 and 20.9, we now consider an internal noise source placed between the two stages of amplification as shown in Fig. 20.10a. Let us treat the signal and the noise independently. To differentiate between the two, boldface letters are used to represent "noise quantities" in Fig. 20.10.

For a signal input E_{in}, Eq. (20.45) dictates the output:

$$E_o = \frac{G}{1 + G\beta} E_{in} \qquad G = G_1 G_2 \qquad (20.48a)$$

For a noise input $\mathbf{N}_{in}$, let us designate $\mathbf{N}_o$ as the noise output. Taking the feedback effect into consideration, we have, between the two stages, (1) noise input $\mathbf{N}_{in}$ and (2) equivalent input $-G_1\mathbf{N}_o\beta$ from the feedback. This amounts to an input $\mathbf{N}_{in} - G_1\mathbf{N}_o\beta$ to the second stage, which yields an output

$$\mathbf{N}_o = G_2(\mathbf{N}_{in} - G_1\mathbf{N}_o\beta)$$

Rearranging the above equation, we find

$$\mathbf{N}_o = \frac{G_2}{1 + G\beta} \mathbf{N}_{in} \qquad G = G_1 G_2 \qquad (20.48b)$$

We shall now take β, G_1, G_2, G, and G' as specified in (20.46) and assume a noise input $\mathbf{N}_{in} = 1$ volt. Let us now try to find the signal/noise ratios for an open-loop arrangement and a closed-loop arrangement for the same signal output $E_o = 100$ volts.

For the *open-loop arrangement* in Fig. 20.10b, $E_o = 100$ volts and $G = 100$ imply that we need an input signal $E_{in} = E_o/G = 1$ volt. From the quantities in the figure, we have the following signal/noise ratio in the output:

$$\frac{\text{Signal}}{\text{Noise}} = \left| \frac{E_o}{\mathbf{N}_o} \right| = 10 \qquad (20.49a)$$

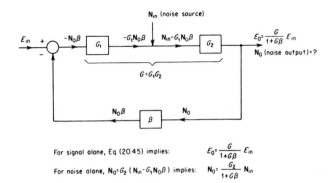

For signal alone, Eq. (20.45) implies: $E_0 = \dfrac{G}{1+G\beta} E_{in}$

For noise alone, $N_0 = G_2 (N_{in} - G_1 N_0 \beta)$ implies: $N_0 = \dfrac{G_2}{1+G\beta} N_{in}$

(a)

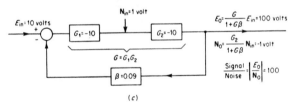

(b)

(c)

FIG. 20.10

For the *closed-loop arrangement* in Fig. 20.10c, $E_o = 100$ volts and $G' = 10$ as obtained in (20.46b) imply that we need an input signal $E_{in} = E_o/G' = 10$ volts. The noise output may be computed with the aid of (20.48b) to be $N_o = -1$ volt. We now have the following signal/noise ratio in the output:

$$\frac{\text{Signal}}{\text{Noise}} = \left|\frac{E_o}{N_o}\right| = 100 \qquad (20.49b)$$

It is always desirable to have a high signal/noise ratio. Comparing Eqs. (20.49). we note an improvement with negative feedback in a closed-loop arrangement.

4. *Reduction in Frequency Distortion as Another Negative-feedback Effect.* It is often desirable to have a uniform frequency characteristic for an amplifier, i.e., to have a relatively constant gain for a wide range of frequency.

Let us consider an amplifier with severe frequency distortion. For example, assume gains

$$G_a = 100 \qquad \text{at frequency } f_a = 1{,}000 \text{ cps} \qquad (20.50a)$$

and

$$G_b = 200 \qquad \text{at frequency } f_b = 2{,}000 \text{ cps} \qquad (20.50b)$$

We shall improve the system with negative feedback. Suppose we use the arrangement in Fig. 20.8a with $\beta = 0.09$. We now find that, with negative feedback,

$$G_a' = \frac{G_a}{1 + G_a\beta} = \frac{100}{1 + 100 \times 0.09} = 10 \qquad \text{at } f_a = 1{,}000 \text{ cps} \quad (20.51a)$$

and

$$G_b' = \frac{G_b}{1 + G_b\beta} = \frac{200}{1 + 200 \times 0.09} = 10.5 \qquad \text{at } f_b = 2{,}000 \text{ cps} \quad (20.51b)$$

Comparing (20.51) with (20.50), we note a marked reduction in frequency distortion with negative feedback.

Classification of a negative-feedback amplifier as a type 0 system. For a "single stage" in the two-stage amplifier in Fig. 20.8a, we have the transfer function or gain†

$$G_1(s) = \begin{cases} \dfrac{-g_m R_L}{1 + \omega_1/s} = \dfrac{-g_m R_L T_1 s}{1 + T_1 s} & \begin{array}{l} s = j\omega; \text{ for } \omega < \omega_0, \\ \text{that is, for } low \ frequencies \end{array} & (20.52a) \\[4mm] \dfrac{-g_m R_L}{1 + s/\omega_2} = \dfrac{-g_m R_L}{1 + T_2 s} & \begin{array}{l} s = j\omega; \text{ for } \omega > \omega_0, \\ \text{that is, for } high \ frequencies \end{array} & (20.52b) \end{cases}$$

where

$$\omega_1 = \frac{1}{R_g C} = \frac{1}{T_1} = \text{low half-power frequency, rad/sec}$$

$$\omega_2 = \frac{1}{R_L C_s} = \frac{1}{T_2} = \text{high half-power frequency, rad/sec}$$

$$\omega_0 = \sqrt{\omega_1 \omega_2} = \text{midband frequency} \qquad (20.53)$$

$g_m, R_L, R_g, C, C_s, \beta$ = circuit parameters in Fig. 20.8a

$$K = (g_m R_L)^2$$

For the two-stage amplifier in Fig. 20.8a, assuming identical stages, that is, $G_1(s) = G_2(s)$, the forward transfer function is $G(s) = G_1(s)G_2(s) = [G_1(s)]^2$; that is,

Forward transfer function:

$$G(s) = \begin{cases} G_{\text{low}}(s) = \dfrac{KT_1^2 s^2}{(1 + T_1 s)^2} & \begin{array}{l} s = j\omega; \text{ for } \omega < \omega_0, \\ \text{that is, for } low \ frequencies \end{array} & (20.54a) \\[4mm] G_{\text{high}}(s) = \dfrac{K}{(1 + T_2 s)^2} & \begin{array}{l} s = j\omega; \text{ for } \omega > \omega_0, \\ \text{that is, for } high \ frequencies \end{array} & (20.54b) \end{cases}$$

where K, T_1, T_2, and ω_0 are defined in (20.53). Comparing (20.54) with (20.19), we see that the feedback amplifier in Fig. 20.8a is a type 0 system according to the general representation of its forward transfer function (20.54).

According to (20.44), we now have

Open-loop transfer function:

$$G(s)H(s) = \begin{cases} G(s)\beta = G_{\text{low}}(s)\beta = \dfrac{K\beta T_1^2 s^2}{(1 + T_1 s)^2} & \begin{array}{l} s = j\omega; \text{ for } \omega < \omega_0, \text{ that} \\ \text{is, for } low \ frequencies \end{array} & (20.55a) \\[4mm] G_{\text{high}}(s)\beta = \dfrac{K\beta}{(1 + T_2 s)^2} & \begin{array}{l} s = j\omega; \text{ for } \omega > \omega_0, \text{ that} \\ \text{is, for } high \ frequencies \end{array} & (20.55b) \end{cases}$$

where K, β, T_1, T_2, and ω_0 are defined in (20.53). Comparing (20.55) with (20.33), we see that the feedback amplifier in Fig. 20.8a is also a type 0 system according to the general representation of its open-loop transfer function.

Remarks about the system characteristics of a negative-feedback amplifier. We have seen that the negative-feedback amplifier in Fig. 20.8a is a type 0 system. We now note:

---

† See MIT Electrical Engineering Staff, "Applied Electronics," pp. 470–485, John Wiley & Sons, Inc., New York, 1946. Approximations made here are $R_L \cong R_{eq}$ and $R_g \cong R_{eq}'$ for practical circuits with $R_L \ll r_p \ll R_g$. Similar expressions are derived in this text [Eqs. (14.41) and (14.31)] for transient study.

1. A constant actuating error, that is, $E_g = E_{in} - \beta E_o$ in our present notation, is needed to produce a constant controlled variable, that is, E_o here; this may be seen in the numerical illustration in Fig. 20.9. This is consistent with (20.28b), which is characteristic of all type 0 systems.

2. For a negative-feedback amplifier, our concern is to have a reliable E_o, and we are assured of it in the above discussion. The need for an actuating error $E_g = E_{in} - \beta E_o$ does not worry us. However, the term *error* is perhaps not appropriate in this case.

Stability of a negative-feedback amplifier as a type 0 system. We shall now investigate the Nyquist diagram $G(j\omega)H(j\omega)$, that is, $G(j\omega)\beta$, for the negative-feedback amplifier in Fig. 20.8a. We merely replace $s = j\omega$ in (20.55), plot the positive-frequency portion of $G(j\omega)\beta$, and then construct the negative-frequency portion through symmetry with respect to the real axis.

Assuming a set of values for the parameters in (20.55), we plot the positive-frequency portion of the Nyquist diagram in Fig. 20.11a. Note that, for the upper half, we used Eq. (20.55a); for the lower half, Eq. (20.55b).

Since the positive-frequency portion of the Nyquist diagram is symmetrical with respect to the real axis, the negative-frequency portion is congruent to the positive-frequency portion. For clarity, we reproduce the positive-frequency portion of the Nyquist diagram in Fig. 20.11a' and the negative-frequency portion in Fig. 20.11b. To trace the complete Nyquist diagram is to trace the heart-shaped figure in Fig. 20.11a twice, once for $\omega = 0^+$ to $\omega = \infty$, and once for $\omega = -\infty$ to $\omega = 0^-$, in the clockwise direction. We now have $P = 0$ [by inspection of (20.55)] and $N = 0$, where P and N are interpreted in (20.34c). According to the Nyquist criterion in Art. 19.3C this is a stable system.

By changing the gain K to a much larger value, we enlarge the Nyquist diagram in Fig. 20.11a proportionally. However, the Nyquist diagram can never encircle -1 for any value of K; the amplifier is therefore *always stable*.

B. Illustration 2: Speed Regulator

A simple speed-regulator system is described in Fig. 20.12a with individual transfer functions taken from Fig. 20.2.

Forward and open-loop transfer functions. We easily recognize that

Forward transfer function:

$$G(s) = \frac{K_a K_m}{1 + T_m s} \tag{20.56a}$$

and

Open-loop transfer function:

$$G(s)H(s) = \frac{K}{1 + T_m s} \qquad K = K_a K_m K_t \tag{20.56b}$$

This speed-regulator system is a type 0 system, as is evident if Eqs. (20.56) are compared with (20.19) and (20.33).

The effects of the feedback loop in this case are much like the effects of negative feedback in the previous illustration, except for notation.

Stability of a speed-regulator system. Replacing $s = j\omega$ in (20.56b), we plot the Nyquist diagram $G(j\omega)H(j\omega)$ of the above-described speed-regulator system in Fig. 20.12b. We now have $P = 0$ [by inspection of (20.56b)] and $N = 0$, where P and N are interpreted in (20.34c). According to the Nyquist criterion in Art. 19.3C, this is a *stable* system.

By changing the gain K to a much larger value, we enlarge the Nyquist diagram in Fig. 20.12b proportionally. However, it can never encircle -1 for any value of K; this speed-regulator system is always stable.

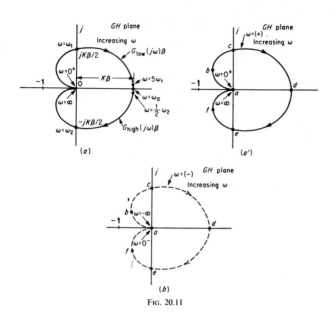

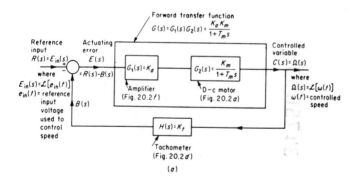

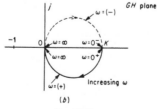

Fig. 20.11

Fig. 20.12

C. Remarks about Type 0 Closed-loop Control Systems

Some characteristics of a type 0 system. From the illustrations we note the following:

1. A type 0 system is suitable for use as a regulator system, for it is less susceptible to component unreliability, disturbances or noises, and distortions than type 1 and 2 systems.

2. A constant actuating error is needed to produce a constant controlled variable. For other characteristics of a type 0 system, see the statements in (20.28).

System stability. From the Nyquist diagrams for type 0 systems in Figs. 20.4, 20.11, and 20.12b, we note:

1. Most type 0 physical systems are stable.

2. If a type 0 system is unstable, we may make it stable by reducing its system gain K or employing compensation, which will be discussed in Chap. 21.

3. The stability problem is not a severe one in type 0 systems as compared with type 1 and 2 systems.